BUTT DIALS

FROM THE

CENTER

OF THE

GALAXY

By Paul Vancleve

BUTT DIALS FROM THE CENTER OF THE GALAXY

TABLE OF CONTENTS

PREFACE

The Universe and its Intrepidness

It takes billions upon billions of years of raw torturous evolution just so in the end, the Universe can feel like Gene Kelly in 'Singing in the Rain'. This is no whimsical thought; for in pursuit of the great comic madness so cherished by all intelligent beings, the Universe must first condense huge million mile in diameter balls of hydrogen into 40 million degree temperatures in space just to begin the process of conscious laughter, which still awaits it countless billions of years into the future. But the Universe is patient. It doesn't mind waiting around a few tens of billions of years to show its feelings. After all, it deals with an ultimate inventory of infinity, which as you can rest assured, will take forever. And in dealing with this ultimate inventory of cosmic possibilities, the Universe will, through its inventory of eternity, explore the fantastic super spectrum of conscious creations it has within it; thus evolving itself into beautiful arrays of cosmic consciousness spanning the depths of forever, experiencing a spectacular variety of life form creations from microbial amoebas to over zealous insurance salesmen, never really knowing what evolution of consciousness it will enter into next. Einstein once said, "God does not play dice with the Universe". This is true; but the Universe also does not know what is going to happen next. Only the Universe as a whole can have these two properties. No single entity within it can have these two aspects. But what does it all really mean? It means the Universe is like a child, constantly discovering itself over and over again into infinity!

WELCOME TO THE MIDDLE OF INFINITY

It's the only place you'll ever be in your life; and also the only place you can ever go. Which begs the question; where and when the hell are we? Are we traveling infinitely fast and don't know it? I mean lets face it; if we are moving with the Earth and the Earth is moving with the Sun and the Sun is moving with the galaxy and the galaxy is moving with the 2 billion other galaxies around it, of which are moving with the 5 trillion other galaxies surrounding them, then at what point does it end?; or does it ever end?; which then begs the question; if we are moving at infinite speed then are we traveling at any speed at all? We may never know because infinity is the only thing that is infinite which gives it kind of a monopoly in the infinity game. This means no one; man nor dolphin can ever truly ever comprehend it. For us; at the end of the day, the only measurable speed that we can ascertain depends on whatever we are traveling relative to. There is no common universal frame of reference for the entire Universe; but conceivably if everything that is moving is part a larger mass that is in itself part of an even larger mass such as galaxies that belong to groups of galaxies which in turn belong to still larger groups of galaxies into infinity, then we could be moving in all directions at once; which means that with an infinity of matter moving in all directions at once then theoretically all motion that is being carried along with it would be cancelled out meaning that nothing has ever moved in the Universe. I can never get a fix on where I am or what direction I am going or how fast I am traveling or if I am traveling at all because everything that is moving me is, itself, being moved by something bigger on up into infinity. This means that we are literally stuck in the middle of infinity without a paddle; which in turn means that everything is at the center of the Universe. This means that Donald Trump really is at the center of the Universe; but so is my cats asshole. It means that everything, in fact, is at the center of the Universe. Confused? Don't sweat it. Infinity does this sort of thing all the time. The point of zero speed is nowhere. It doesn't exist. There is always something slower or there is always something faster in all directions infinitely. I take it back I'd rather be stuck out in the middle of infinity than nowhere; for nowhere is relative to nothing but at least the Universe is always relative to something. That's the thing about the Universe! Even if you reach the end of infinity you're still stuck right smack in the middle of it; and if there anything worse than being stuck out in the middle of nowhere it is being stuck out in the middle of nowhere in the middle of infinity; talk about being up a wormhole without a fusion engine!

MASS COMMUNICATION BIRTH

Welcome to Mass Communication Super Hell Second on the Planet Earth representing nothing less than a cosmic event in the evolution of the Galaxy. I always like to start my books off by telling people what time it is; in terms of evolution, that is. It is a wonderful time to be alive provided that you have no idea what the hell is going on; and only in the last cosmic second has the insanity exploded in existence. As we can see below; for millions of years things remained the same; relatively unchanged; boring mindless rock formations; occasionally the wind would blow to spice things up;

20 million years ago

20 minutes ago

then suddenly, in an instant, we've got *over zealous conmen trying to email lonely romantics with photoshoped pictures of fake people in order to scam money by shipping merchandise overseas for the promise of a phony percentage in order to trade goods for arms to sell to radical militants who just wanna to drive down big oil production in order to reclaim the holy land for the purposes of starting a revolution in order to create an army in order to make war against opposing forces in neighboring lands* Did you get all that?

And only in the last instant did the party start; and don't think that the Gods didn't notice either. "They're watching; I know"; as Ralph Kramden might say. And why not?; this is where the action is. We have practically everything there is, trapped in a swirling tornado of death and orgasm.

With all the chaos and shenanigans that are happening at Mass Communication Birth we may very well be the 'Breaking Bad' of galactic television of which the gods never miss an episode.

Mass Communication Birth is a cesspool of exploding information happening at the speed of light documenting every ugly beautiful aspect of humanity that can ever be imagined. From mindless global media super gossip to whole libraries that you can carry in your wallet; it has been multiple digital evolutions exploding from multiple analog evolutions exploding from multiple technological innovations dating hundreds of years back; pop, pop, pop; all happening around us without warning. And embedded within it all is a multitude of cosmic occurrence so intriguing that the Gods themselves may actually be binge watching.

There is, of course, our threat to ourselves of potential thermonuclear destruction occurring in the same second; But it makes sense; I mean how could we have any appreciation at all for all the crazy nutty imaginations that are happening all around us if the planet wasn't under a constant threat of immediate attack by a hydrogen bomb surplus with another eight thousand insurance bombs waiting in backup just to be sure. As a remedy of hope though, we also have our birth of Technological Construction in Outer Space occurring in the same second so that in case that whole nuclear thing goes south, the human race can live on in deep space mining the hell out of whatever moves.

This might come in handy too because Mass Communication Birth also exists inside the Earths 6th Great Mass Extinction Event. But unlike previous mass extinction events this one is a technological mass extinction which means that a wide variety of life forms will be annihilated in a multitude of different ways creating a multitude of different evolutionary directions for the survivors to evolve in. But the good news is, mass extinctions are about the living; not the dead; so they are not total extinctions but instead are about whatever is clever enough to survive and pass on *only* their genes into the future. But Mass Communication Birth has a fun side to it too. Supersonic travel has enabled globalized access to people and product around the planet in near immediate time relatively speaking; and if you get bored on the flight we also have a birth of pornographic exchange that is also occurring at the speed of light in the same cosmic second; and this, in itself, may represent a further bonding agent of mammalian evolution to defend against potential thermonuclear destruction of humanity. Drop that on

Mother the next time she throws away your best porno mag.

All of this adds up to what amounts to a super second in Evolution; a super cosmic second that stretches 30,000 years across space and time; from the birth of civilization 20,000 years ago to 10,000 years into the future, Mass Communication Birth represents the cosmic collision of high speed multiples of information exchange that could not be explained to anyone who hasn't experienced it;
and as with any increase in evolutionary complexity, Mass Communication Birth brings about new mechanisms in evolution not previously imagined.

Why we will evolve into cosmic silliness to avoid thermonuclear holocaust

In our long evolution, only in the last billionth of a second did we become consciously aware of our existence. Now we are on the verge of our birth of technological construction in outer space. It is the beginning of an infinite exploration into forever. And it is right now, at our birth of technological construction in outer space that we have discovered our birth of mass communication technology, creating, perhaps, that of cosmic episode in the evolution of the Milky Way Galaxy. Why? Because at this second in our evolution we have discovered the ability to transmit audio and motion visual information around the planet at the speed of light, allowing fantastic displays of knowledge to come at us so quickly from all directions, that we can now make vast arrays of new discoveries every second. It is the evolution of global exchange at the speed of light, allowing the accumulation of literally hundreds of billions of times more knowledge, literally millions of times quicker in less than a second (*By the way, the internet only weighs a half an ounce*). And this may very well be a noticeable event; at least in terms of whatever Gods may be ease dropping in on us. It may, in fact, be a super event in the evolution of the Milky Way...a dynamic milestone in the evolution of life; the absolute breakneck race of comic insanity versus potential thermonuclear destruction, both occurring in perfect synchronization with one another to produce the maximum evolution of human and dolphin intelligence. In short, the Gods just might actually give a damn; and who knows, we might even be a galactic reality show hit of some kind. I stay in makeup at all times now. And the comforting thing about all this mass annihilation and death is that the Universe does not actually want to destroy us, but instead, just wants us to evolve into cosmic silliness to avoid total annihilation of the Earth. So relax! But achieving 'cosmic silliness' will not be pretty. It never is! In fact, it is a road paved with genocide and barbarism. I'd like to say we will have a happy ending, but unfortunately, Steven Spielberg is not running the Universe. So let's just say we need to come as close as possible to nuclear annihilation in order to achieve the highest level of comic insanity possible; and believe it or not, the Universe, which is conversely trying to stab us in the back, wants us to succeed; and we will. In time our children will move on to colonize the stars and eventually in our super distant evolution, journey inward towards the Center of the Galaxy to evolve into the Super Galactic Network of Fantastic Life Forms evolving faster and faster into one another generating the massive buildup of the evolution of the Galaxy. But first we have to get the F. off of this planet intact, so go Elon Musk!

Currently we are the product of 10 billion years of stellar evolution that is currently undergoing the spectacular conversion of hydrogen into orgasm. At our present evolution we are a technological species just entering into our birth of digital chip evolution at speeds approaching 10 gigahertz. We have spawned the birth of a technological consciousness at our birth of mass communication, although we are unsure of its ultimate direction in terms of cosmic evolution. But that is the way of the Universe. It will always be ahead of us by natural occurrence. No matter how advanced we get, the Universe will always be there to kill those who it feels no longer provide sexual inspiration in terms of evolution. It is the ultimate battle of 'horny vs. evil'; and behind us are literally 100's of billions of life forms competing for evolution dominance on the environments of the Planet Earth. And each must endure the complete unsympathetic happenings of the Universe. And although they do not know it, they are competing for the right to exist in conscious ecstasy deep in the depths of space. Little do they know how weird it gets though. A hundred years ago we were riding horses through the wild unknown. Now, only a hundred years after swimming knee deep in horse shit, we are regularly sending robots to Jupiter, Saturn and beyond, completing our first reconnaissance of the Solar System. In less than a thousand years we may have technologically evolved beyond what we can perceive now. Our emergence beyond the speed of light soon after may allow a view from outside, looking in, at this Atomic Dimension that we were created in. But first, let us understand who we are and what it is that we are supposed to kill in terms of evolution.

EVOLUTION THEORY GONE OVERBOARD

That would be me!

Warning: If you are heavy into religion, viewer discretion is advised!

There is no bigger serial killer in the Universe than the Universe itself. It wants to kill practically everything there is, and for the rest, it wants only sex. And it is the entire Universe that wants this; right down to the photon. The heavens are filled with the cosmic machinery of evolution; even the two gas giants; Jupiter and Saturn, that just spin around and around for billions of years and do nothing but be big, assume a massive role in the evolution of life on Earth. It is they, whose solar orbital positions around the Sun, determine gravitationally, the direction and timing of the asteroids and comets from the Ort cloud and Kuiper belt, that will ultimately strike the Earth and cause the mass extinctions, which will then determine what species of life forms will die and what species of life forms shall live to pass on *only* their genes into the future. We've even seen Jupiter eat comets for dinner as with Comet Shoemaker/Levy 9 in 1994. Other times it only alters their course, perhaps towards or away from the Earth. Jupiter and Saturn are like Gambino and Sammy the Bull; They scare the living shit out of any comet that that even thinks about a close approach. But more than likely John and Sammy just want them to work for them which usually means performing a hit on the Earth at some future precise time. Whether friend or foe, Jupiter has, no doubt, affected the direction and speed of many kilometer sized bullets, either away from the Earth or into the Earth for just the right bombardment of Earth, to assist in creating a balanced evolution. And it doesn't take much of a hit either. Try to imagine an asteroid collision with the Earth like this. Picture a bowl of water sitting in an empty room perfectly still. It shows no apparent movement. It represents the Earth. Then somebody drops a single grain of pepper into it. If you look at the pepper when it hits the water you can see the ripples expand out from the center of the bowl. Those ripples are equal to waves thousands of feet tall on the Earth. It is mass destruction of an otherwise calm environment. It's called EVOLUTION and it is a basic foundation for the evolution of consciousness; and there is nothing in the Universe that is not IT. It is the total indifference of particles colliding in perfect math to allow the Universe to happen blindly. This allows evolution to create chaos for only the best to survive. Evolution occurs in ways that prevent even our technologically constructed environment from defending us against the constant chaotic change of the evolution environment in perpetual motion. We are designed to change willingly at only certain rates but evolution happens much faster and as a result we are always adapting to stay afloat.

The Earth/Moon system itself could also be said to be an evolution machine of rotating gears designed by no one, that stabilizes the rotation of the Earth to allow for the proper cycles of the seasons to take place. This is absolutely crucial for the development of the evolution of life on Earth. It allows Evolution to be a cyclic phenomenon and not just something that occurs in never ending chaos. Orbital Stability for seasonal change creates an environment of stable evolutionary change. The Moon allows for a stable Earth orbit which allows the Earth to receive the proper amount of heat energy from the Sun to sustain long term evolution. It is also responsible for the stability of the tides and weather as well. It is the Moon that may have even aided in the delivery of aquatic based life to the land, through massive tides, for future land based evolution nearly 400 million years ago. What's the point you ask? The point is that the Moon, which singlehandedly stabilizes the Earth's rotation for the proper evolution of life on the planet, is the only moon in the Solar System that performs this task. Mercury and Venus have no moons and would both be too hot anyway for the evolution of life. Mars is too cold and only has two small captured asteroids for moons in which neither, in no way at all, comes close to stabilizing the rotation of the planet Mars. Jupiter, Saturn, Uranus, and Neptune have more than 100 known moons yet none come anywhere close to the size required to affect any of these gas giants orbital rotation. Isn't is funny how only the Earth has a moon of just the right size and distance to stabilize the rotation of the Earth for the evolution of life while all the other planets, which don't need a stabilizing satellite, don't even have one by accident. But as I said before, all of the other planets still play a crucial role in the evolution of life by gravitationally affecting the comets and asteroids that are, in fact, flying around the solar system like bullets in a war zone. It is these comets and asteroids that have, and will again, collide with the Earth to create the mass extinctions that will allow only those who survive to pass on their genes for the future creation of new species. Too many collisions with the Earth and all life could be wiped out or only microbial life survives every time; too little collision and you slow down the rate of change of evolution by allowing species who have become too specialized to continue breeding. Evolution would still take place, of course, but it would be slowed down by literally hundreds of millions of years. Mass extinctions are nice for evolution because they wipe out entire species of life in a matter of days or maybe even seconds. When the Universe is done with you then it is done with you; and there is almost never a farewell party to say goodbye. It just cuts straight to the chase; that way the Universe doesn't have to wait around for nature or climate change to direct evolution; and what is the result in the end?; more intelligent life is able to be created throughout the life of the Earth.

Proving that evolution is real is not difficult for anyone who is willing to listen. The above illustration utilizes a species in transformation as proof. As we can see above dolphins have finger bones in their flippers. Why would finger bones exist in a flipper that is simply used to steer currents? The answer is because the dolphin was once a land animal tens of millions of years into the past where it needed a certain amount of dexterity to survive on land. As it began to move back into the water in terms of evolution the old adage of it's former appendages worked to keep the flipper solid but having finger bones to support a flipper is not necessary. The finger bones are simply left over genetic code of what the mammal was in the past.

Fun with Mass Extinction

However in the particular mass extinction of which humanities Mass Communication Birth is embedded, it is a new kind of mass extinction that is taking place.

Instead of an asteroid impact or volcanism or naturally occurring climate change we get to experience a new forms of mass extinction that I refer to as a technological mass extinction. To understand this type of mass extinction we must first understand what happens during the run of the mill cliché mass extinction that killed the dinosaurs. Here we had an asteroid impact from an asteroid 6 miles across impacting in the Yucatan and kicking up enough dust to block out sunlight for years to come. This, of course, diminished plant life which then worked its way up the food chain causing whatever had survived to get smaller. In other words everybody more or less had to deal with the same issues. But a technological mass extinction is different. This is a mass extinction that is initiated by conscious beings in a multitude of different ways around the planet. You may have big cats poached to death in one area while frogs die of atmospheric changes in another area while fish are decimated from consuming pollution in yet another area. Unless the species in question goes totally extinct then it will be the survivors who will then pass on only their genes into the future. But the method of death determines the direction of evolution that the surviving species will take. So if Big Cats are being trophy shot into near extinction then only those cats that have a higher intelligence to determine the immediate danger of the human hunter may survive. Likewise, if frogs are dying from absorption of pollutants through the skin then only that small percentage of frogs that are immune or can absorb more pollutants through skin may survive. As we can see the method of death determines the direction of evolution. And in the case of this technological mass extinction, the direction of evolution is not only technological but wide spread in terms of its variety.

Human beings have initiated these mass extinctions on animal and plant life but the mass extinction scenarios that they have created are now beginning to engulf themselves as well; which reminds me; I have to check and make sure that my voter registration card is up to date for the 2020 presidential election.

Alright, to be fair nobody can really state that the Universe exists in any particular state such as an ongoing state of evolution reasoning; 'evolution reasoning' meaning that all the events that are happening around us are taking place in a natural occurrence of evolutionary cause and effect which are embedded within the law of physics. But whatever state it does exists in, evolution is certainly involved. In terms of the Universe being, in fact, pure evolution, in which even inanimate non life forms such as the other planets of the solar system are part of the evolutionary process; the theory that the Universe is pure evolution **does not** fall apart as we move in this direction of inanimacy.

If one accepts the idea of Evolution at least having a role in the workings of this Universe then one is free to move on to ask more in depth questions about the forms that evolution might occur in. By this I mean the great patterns of the Universe that generate themselves again and again throughout infinity into forever. These are the great blues symphonies of the Universe occurring in harmonic occurrence; perhaps strung through the motions of the planets and stars in harmonic purpose; they are the sequence of the 5, 7 and 12 combination. The harmonic notation, in a very real sense, is the music of the Universe. In fact, they actually occur naturally in musical notation in the form of octaves which, of course, are embedded within the laws of physics.

Intelligent design for the purposes of evolutionary occurrence

Don't ever think that intelligent design and evolution can't coexist in the same particle dynamics. Evolution is a fact; if you don't believe it then don't waste your time getting that flu shot to protect against the mutation of the H1N1 flu virus for next years flu season. But the creation of the particle dynamics that occurs in evolution doesn't necessarily have to be a natural formation. Already today we have the genetic manipulation of crop seeds for the eventual consumption of the unwitting consumer. Atomic particle manipulation for the creation an atomic dimension such as the one that we live in is certainly as real a possibility as anything. Matter pouring into this dimension to ultimately allow stellar evolution to power cuteness for the purposes of evolutionary selection may very well be an engineering marvel of the particle dynamical structure of the atom. Our atomic makeup may very well be the pride and joy creation of a farmer god who then took a handful of that atomic seed and spread it out into space for the purposes of the ignition of consciousness. And like the seeds planted by a farmer, that god may have a definite idea of what that seed will do and become but only to a point. I mean after all; who wants to know everything that is going to happen down to infinity. Gods, no doubt, like to be surprised as well.

An exactitude of the laws of physics should imply an exactitude of the laws of events

When I get up in the morning I expect gravity to not only be turned on but to function exactly as it did yesterday. And so everyone agrees that the laws of physics as we experience them on Earth are absolutely reliable in terms of their exactitude. I mean, if you're like an electrician or a tight rope walker; you depend on these things. Yet in spite of this many physicists assume that the Universe must occur in chaos. Why?; because to imagine every particle in the Sun behaving exactly as it should in some crazy cosmically oriented conspiracy is not provable and is also very difficult to imagine because there is so dammed many atoms. That's one of the complaints that I've always had about this Universe; there's too many atoms...anyway...

Imagine a pool table with the balls in triangular position ready to be broke.

Then the break; the balls go everywhere.

Now we rewind the clock back to the exact same scenario. It is the exact same scenario down to INFINITY. In other words; everything is infinitely the same. I shoot and again the same thing happens again infinitely exactly the same as before. It has to be exactly the same because there is no difference in existence in the Universe to make it come out differently. But the chaos theorist will tell you that it will always be different. But the fact is that in order for it to happen differently then there must be some DIFFERENCE between the two scenarios. But this scenario also means something else; it also means that there is only one future that can happen; one and only one! These chaos theorists have had a free lunch for long enough but now the party is over.

Now if you ask me; all of this points to an exactitude of evolution occurrence because the sole purpose of the hydrogen atom is to achieve orgasm but don't quote me on that; and yes, I do actually believe that; and I admit, of course, that it is wildly subjective; unless, of course, you happen to be having an orgasm at the time; then its easier to believe; but the scenario involving the pool table is straight forward LOGIC. Do the Math!

Time for more acid

And because the Universe may be a big ol horny ball of pure evolution occurring into fractal infinity; strange harmonious events, millions or even billions of years in the making, may occur in our everyday lives as seemingly mundane everyday events. What may begin with an ape picking berries off a tree 20 million years ago may directly lead to someone typing in a debit number to buy a pound of grapes at a local Walmart checkout five million years later. They would be harmonic of one another in terms of the laws of physics. It would be two events, both seemingly insignificant, that are connected by millions of years of direct harmonic event notation; one event leading to another; and connecting the two is a precise evolutionary sequence of events millions of years old that is exact down to an infinite level. But why, you ask, would these events maybe be connected? The answer is that they are repeating in terms of harmonic evolution; and in a very real sense they are the same thing. Mark Twain once stated *'Nature does not repeat, but it does rhyme'*. An Australopithecus ape picking berries is gathering food. A homo sapiens (modern human) typing in a debit number at a Walmart is also gathering food. The modern human is just doing it in a way considerably more complex than the way the Australopithecus did it. The irony is that the Australopithecus is probably eating a more balanced meal; but anyway…they are still achieving the same result; food for survival! The events repeat like octaves on a scale. But they repeat in terms of Evolution as well! These everyday mundane events are directly tied to events millions of years old and occur in apparent complete mundaneness. In other words you won't see the Universe throwing any ticker tape parades just because it reached some harmonic point in evolution that took a billion years to achieve. To the Universe the death of the President of the United States and the death of a shrew are equal *in* importance *(in the case of Trump the shrew might win)*. The Universe doesn't stop to look in amazement at what just happened. It just keeps occurring in mindless evolution without regard for anything other than the complete transformation of hydrogen into orgasm through stellar evolution. The reality of the blind, naturally occurring math of the Universe to achieve orgasm is utterly merciless in its indifference to what anything in the Universe might actually want. It pauses at no point to celebrate or to mourn. It simply plows onward and into the forever to create into existence every possible creation imaginable in the naturally occurring construction of the hydrogen atom. And most likely the evolutionary reasoning that may be happening all around us and to us has far deeper meaning in its agenda, which by the way, may be infinitely complex to keep consciousness evolving into infinity in its never ending attempts to understand it. Make no mistake about it; the Universe has an agenda. For all we know the Universe may have evolved human beings just so they could develop nuclear weapons and destroy themselves so that the cockroach can evolve from the radiation to survive the massive cosmic ray bombardment of deep space so they could make their way towards the Center of the Galaxy to attend some perverted cockroach orgy 4 billion years from now; and don't kid yourself; if the Universe has anything; its infinite patience for infinite perversion.

Do planets with promising technological forms ever go extinct by extra planetary forces ie. asteroid or nuclear war leaving absolutely nothing behind.

I believe we live in an exactitude Dimension. This means that the Universe has apparent intent...at least in my view. That intent is pure evolution verb ... evolution action, in motion to produce something that survives and is better than it was. That being said then I must ask the question. Would the Universe bother to evolve humanity through 4 billion years of evolution with all of it's technological potential, not only for humanity but also every other intelligence that might benefit from the use of some whatchamacallit that might transform their existence, to destroy everything and basically sign the planet over to the cockroaches. History, so far, has suggested, at least on this planet, that the forces of evil such as asteroid impacts and thermonuclear war, happen only in the name of the future evolution of the survivors. Ha, famous last words, I know. Do I dare say that we may have a safe thing going here and that the deaths of almost 7 billion people with a few lucky survivors is basically the norm for this atomic dimension. No, I won't say that even though I just did. I have this overwhelming feeling that if I disrespect the Universe by making an assumption as to our absolute survival, as though it's already in the bag, that the Universe might feel the need to prove a point. That being, of course, that it's not afraid to wipe out the entire human race just because somebody smarted off. Most likely, though, it probably wouldn't destroy all of humanity, just me.

Can humanity exist in a velocity mass change faster than this matter.

In other words: can we exist in a physics different than the physics that makes us up? It could be that the hydrogen atom can exist in terms of the way that it interacts in other dimensions yet be complex in a way that allows it to behave in an ultra dimensional way far more complex that the traditional hydrogen atom does in this atomic dimension. In other words, as the transformation of matter into another atomic dimension takes place, the matter gets to keep original characteristics as it becomes more than what it was in it's new dimensional atomic makeup as induced from velocity.

Everything is a perfect storm

Don't worry about that whole cockroach thing; just forget I brought it up. Let me now give you a new perception in terms of the viewing events; the mysterious 'perfect storm' concept that may very well have stuck its nose in every event that ever happened in the history of the Universe. Lets take the sinking of the Titanic. What does it really take to sink the Titanic? Evidently not much! But seriously, if one really looks at the Titanic and understands it in detail one might begin to see the evolution of events in a whole new light! 1500 people died in the Titanic disaster. And if one really looks at the events surrounding it, one might get the idea that if any one of a number of events **that were not connected to one another** didn't happen then the tragedy wouldn't have occurred or maybe a 1000 people who were killed would have survived.
For instance;

if the Titanic had not changed course;

if the iceberg had not had a longer than normal lifespan and drift rate;

if Bruce Ismay had not strongly suggested that Captain Smith increase speed and ignore iceberg warnings to make port sooner;

if the titanic radio operators hadn't ignored the Californian's warning of icebergs in the area;

if the water had not been so calm maybe the lookouts would have seen the water break around the iceberg sooner;

if First Officer Blair had not taken the keys to the binocular case in his departure from the ship;

if Murdock had not ordered a "hard to port" at just the moment he did the tragedy would have never occurred because the ship would either have rammed the iceberg directly, in which case it would have never sunk, or missed it completely;

if the ship had enough lifeboats (instead of half) to fill then all the passengers and crew would have been saved;

if the life boats they had would have been filled to capacity hundreds would have been saved;

if the radio operator of the Californian hadn't fallen asleep and the equipment shut off while the Titanic was sending it's CQD distress signal;

if the water tight bulk heads had extended all the way

to the top of the ship then the water would have never spilled over and flooded the ship.

As you can see it takes a lot of work and cooperation to create a tragedy. Everyone must work together like a well oiled machine to make it happen; cooperation, teamwork; that's what makes it all come together. But what's interesting about these events is that if you remove any one of them then you basically eliminate the tragedy completely suggesting that what is actually happening is a very well orchestrated design by something beyond our earthly understanding. Remember, none of these things were done by human beings to consciously create a tragedy. And none of these events are really related directly to one another in terms of one leading to another; they just all came together in perfect synchronicity to kill 1500 people. Remember also that the iceberg that the Titanic struck may have been thousands of years old. But that is the way of events in this Universe; events coming out of nowhere, millions or perhaps billions of years in the making, that come into creation in perfect unity in space and time with one another to create some new event that could only happen if all of these events occurred in past synchronicity. It is a perfect storm but it also suggests that everything that ever happened in the Universe is also a perfect storm of events.

Three sperm whales examine the sunken bow of the Titanic.

THE POSSIBILITY OF ACTUALLY PARTYING WITH ALIENS

As has been stated by Carl Sagan; the chances of two alien civilizations encountering one another at approximately the same technological level would be astronomically remote. Think of it like this; imagine you have two interstellar discs of roughly the same mass, composition and density evolving parallel with one another over a period of 4.6 billion years producing technological civilizations that have essentially evolved as twins with respect to one another. If at the end of that 4.6 billion years, one civilization is even .0046 ahead of the other then it will essentially be more than a million years ahead of the other.

So just how weird does it get?

Human beings always like to imagine aliens as being these freaky looking little bipeds that jump into their spaceships, leave their home planet and head for the Earth without a whole lot of history involved. But the reality of the complexity of alien encounters as a result of millions of years of evolutionary brewing could be somewhat more complex. More than likely any alien that has achieved mass communication birth will be part of an intelligence network. On this planet alone, we are looking at the blending of at least three intelligences; human intelligence, dolphin intelligence and artificial intelligence. And we haven't even started interacting with Europa or whatever may be living underneath all these ice worlds orbiting the big

Jovian planets. But no doubt there has been physical interaction with the Earth and some form of alien intelligence in the last 100 million years. And to be honest with you, I wouldn't be surprised to learn that there are actually creatures of the Earth that have been kidnapped and lifted into space by extraterrestrials for journeys inward towards the Center of the Milky Way Galaxy, and that they have, in fact, been making those journeys for millions of years. What the fate of those creatures would be is beyond my pathetic little 'Gilligan's Island' watching brain to comprehend but because of the vastness of both space and time there are a multitude of spectacular evolutions that could have conceivably taken place.

Dr. T-Rex, PhD

Advance genetic manipulations could very well produce new super versions of virtually any species.

They could be making Frankenaliens up there of the most bizarre kind. And many of those creatures could have gone on to successful careers in the galactic insanity that ensued them. I mean lets face it; enough time has passed that there could be anything up there. The last tyrannosaurs lived 66 million years ago. Most people don't understand just how much time that really is. Tyrannosaurus rex could have been lifted from this planet 80 million years ago by intelligences with nano genetic technologies never even hinted at even in our own science fiction. Galactic evolution of life forms of a magnitude beyond human comprehension has no doubt happened. And something like a t-rex could easily have been absorbed into a galactic network of unimaginable complexity to then become something that we as human beings would have no life reference to comprehend. They could have undergone spectacular genetic manipulation both physiologically and cognitively to the point where they could all be super multiformational PhD titans by now. They could be practicing medicine, engineering, law, or God knows what other unimaginable professions. They, in fact, could be gods by now. But it is important to understand that they would, no doubt, have interblended with both naturally forming physiological aliens as well as with artificial intelligences. Yes, 70 million years ago T rexes could have been operating computers in deep space, surfing t rex porn, and looking for that special someone on the intergalactic internet. Then finally they visit the Earth in present day, looking nothing like what we saw in Jurassic Park, essentially with god like intelligence performing God knows what experiments on the Earth. There is more than enough time for this sort of interaction to take place. The important thing to remember is that if we have a close encounter with aliens, then it will probably be only a tiny fraction of a network of alien intelligences.

Poodle gods

Theoretically any creature could be subject to these kinds of interaction now or in the future. There could eventually be miniature poodles constructing whole worlds for travel around black holes into other atomic dimensions beyond our own. They would, of course, no longer resemble their original physiological state of being having been genetically advanced by technologies beyond our ability to imagine over vast expanses of space and time; so you can put that little image of Kittyboo as a kitty fine tuning that Dyson sphere that surrounds an entire solar system right out of your mind…if possible.

Let's party Dude

Now I realize that some of you have this notion in the back of your head that it would be cool to be kidnapped by aliens, taken aboard their starship and whisked off to the stars on some spectacular cosmic adventure; like they're going to actually let you hang out with em' or something. But the reality of it is that if you were actually kidnapped by aliens then you would have about as many rights and amenities as, maybe, a death row inmate; minus the one hour basketball practice every day. I mean the food would be shitty. It would probably smell of a horrible sterilization. You would have nothing to call your own; actually that may not be true; you probably would have one thing that you could call your own and that would be your very own pot to piss in. Sanitation would probably be an issue for any life form hosting other life forms; so having a pot to piss in could be considered a logical assumption. And you would want to take care of that pot too because that pot could basically be your whole social life; so you would want to do things with your pot other than just piss in it. You would want to be nice to your pot; make plans with your pot; comfort your pot; talk to your pot; open up to your pot because it would be about the only thing that you would have in your life to call your own. It would be your big thing in life.

The author demonstrates a new respect for the term 'having a pot to piss in' when imagining how horrific it could be for anything unfortunate enough to be designated as an 'unwitting ambassador' to the stars.

But believe it or not, if you were kidnapped by aliens and taken aboard their starship, your chances of escape would actually not be that bad. It would be about on par with one of these fish (next page) actually jumping out of its tank;

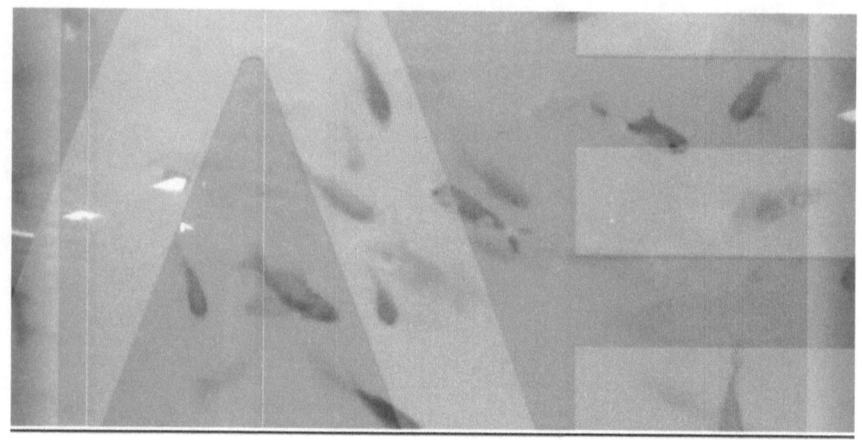

Landing on the ground; then flapping its way out into the world

Learning to drive an automobile;

Finding a good job; getting credit;

eventually getting married; and, of course, the whole
time nobody notices that it's really just a little fish.

Now to be fair you might not end up in a death row
situation as mentioned above; you might end up on a
farm of some kind with various species of roommate
from all over the Earth because there is still a
microbiological contamination concern. But even then

you could end up with a roommate situation you never imagined possible. I mean imagine sharing a room with Otto of the Elephant. If all this sounds crazy then take a look at our zoological establishments; many times there is no real consideration for the mental contentment of the creatures being confined other than the immediate needs for life sustainment. And as for the aliens; you'll never know what their true intentions really are. You may be a part of a zoo type situation where they're just going to study you and other aliens are going to come to see you, or, on the other hand, maybe they're preparing you; maybe you're being basted or something. You could end up being the soufflé cooked by some alien chick that can't cook that everybody's trying to toss it to the side when nobody's looking.

The point is, you would have very little interaction with them whatsoever other than just as means of observation and genetic manipulation. Aliens are going to have lives too and most likely will not be sensitive to the emotional needs of a captive creature that they have only a partial understanding of. Don't get me wrong …they may be very affectionate but as we have seen in our containment of life here on Earth … affection and torture often go hand in hand.

How I would handle it if I were an alien

But you know what? Let's break this down logically. Now we on the Earth are on the verge of a nano technological revolution. I mean let's face it, in less than 1000 years or so we're gonna have video cameras and sound recorders and mass spectrometers that will be small enough to fit on the head of a pin. I mean, if you're a pervert it will be like a dream come true. You will probably be able to get HBO on your toenails.

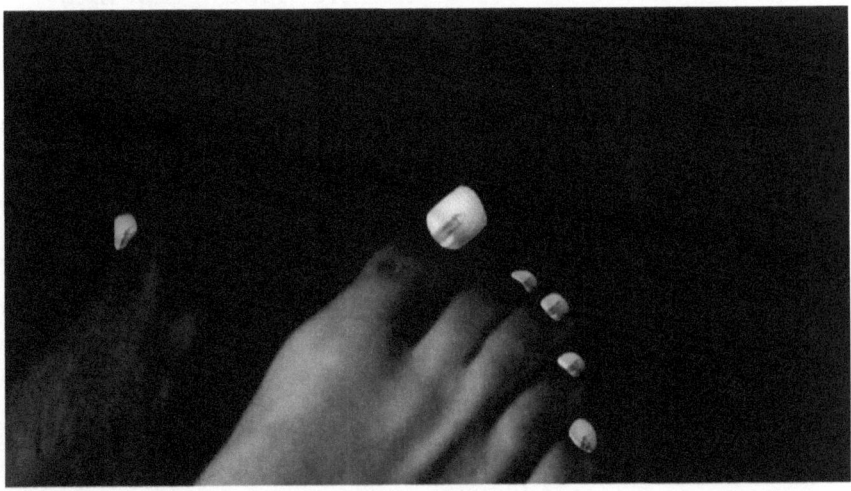

So it makes sense that if an alien species can travel from one star to another then they are going to have at least that level of technology and probably a lot more.

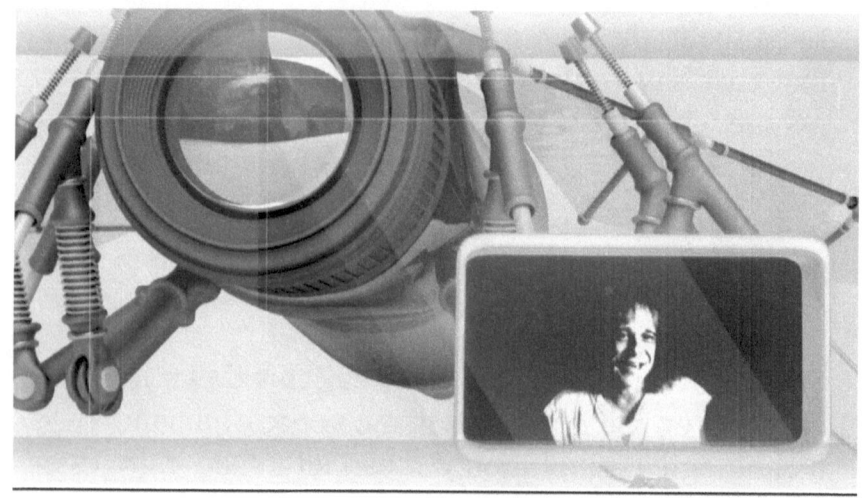

So it doesn't really make any sense why they would take the risk; particularly the microbial risk; to come down here physically when all they would have to do is build some bug sized robot fly looking thing and send it down to collect all the information that they wanted; and as long as nobody stepped on it or sprayed bug spray on it, then it should be fine. It could just shoot its information back up to some golf ball sized mother ship and they would have it. Unless, of course, they're just kinky bastards.

Aliens and their Obsession with Ass

I've never really been much of a grey guy. To me it never really made a whole lot of sense why an alien civilization would travel thousands of light-years across the galaxy just to do the things they're supposedly doing to us. I'm not trying to make a cheap joke here; but in all seriousness, it always seems to involve some sort of butt probing. They never seem to miss an opportunity to do this; and I'm not really sure that all the people who claim it happened to them are really all that bothered by it either. I guess if you were an alien civilization out and about collecting specimens for research then this practice might be necessary to a certain extent. But I have a feeling that the people telling these stories may also have, coincidently, a particular fetish for this type of activity. What a coincidence. I've personally never been convinced that people are being kidnapped by little gray aliens; Still, if I'm out on the road late at night and I see a strange light coming at me and I didn't know what it was, I'd probably would put a butt plug. I wouldn't take a chance. I don't leave home without it.

It's Called the 'WOW' Signal for a Reason

Despite what many people may think; there are credible alien encounters of various sorts that have happened to humanity despite the fact that 99.9 percent of alien encounters are of the crapola classification documentation.

On August 15, 1977 the Earth picked up the 'Wow' Signal. This is a signal that may very well have originated in deep space. It was picked up in Ohio and lasted about 72 seconds and so far has defied all attempts to be explained as a possible terrestrial source. My opinion of what the Wow Signal was, was that it was probably a piece of software update between a planet and a starship or a starship and a starship and somehow the Earth got in way. It may very well be that we just picked up a little piece of someone's communication or upgrade as it was being transmitted to some distant starship. I say this because I would imagine that starships probably receive software upgrades on a regular basis. Many times these upgrade are thousands of years old by the time they reach their destination.

But there could also be a far more perverted answer as well. You've heard of the ancient cosmic background radiation emanating from the Big Bang Explosion that supposedly created the Universe, right? But there may also be an Ancient Cosmic Background Porn as well, emanating from all directions; porn thousands or even millions of years old that has been transmitted by erotic exhibitionist aliens long, long ago. It could be porn so alien we might not even know that we are looking at porn.

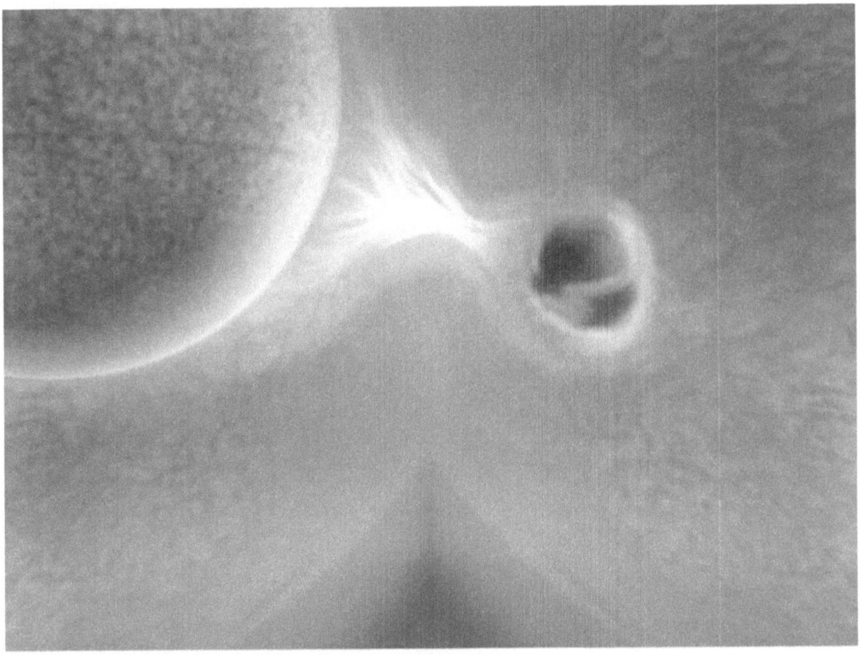

Who knows what alien porn might actually look like?
Above we have an alien cutie ball of some kind being
fed an erotic transmission while being approached by
a gigantic anal anomaly for oral integration.

Previous: Here we have a XXX image of an alien posing for an erotic photograph as it engages in intercourse with an entire planet.

Finally we have a bio robotic cyborg shown in a transmission to it's home planet enjoying a little down on a world where clearly the problem of erectile dysfunction has been solved.

I mean why not? A double digit percentage of homo sapiens internet usage and transmission is actually of a pornographic nature. The Wow Signal could very well be a piece of ass that we picked up accidently. There has to be fragments of porn everywhere in the Universe by now. I'd like to see what it looks like but then again I'm kinda of a voyeur. It would, of course, being far younger than any cosmic background radiation that supposedly is being emanated from the Big Bang Explosion that created the Universe.

Most extraterrestrial candidate signals like the 'Wow' signal usually have a dubious final death knell in that they are ultimately revealed to be of a terrestrial source of some kind because a satellite emitted a weird signal or somebody tried to flush too much Charmin down the Air Force One toilet or something really ridiculous that destroys the cosmic mystery of the whole thing. But so far the Wow Signal has survived as a possible extraterrestrial source. But for how long?; how long before some evil genius comes along and looks at it too hard and destroys the whole thing by finding out the truth of what's really behind it? There's something really depressing in the fact that the best possible example of an extraterrestrial source could be destroyed by a heartless, mindless PH.D competent enough to actually figure it out. That's why I've always believed that the Wow Signal should be examined and studied by only the world's worst scientists because I'm much more comfortable in believing that it was a legitimate extraterrestrial source and are not in need any heartless astrophysicists to come along and prove that this signal was really just a jumbled, reflected transmission of a 'Starsky and Hutch' episode. So that's why I'm going to go out on a limb here and suggest that maybe this investigation should be handed over to only our worst scientists. I mean my point is; we've already conducted a thorough search for what this signal may be and came up empty. Why press our luck? Let's quit while we are ahead and hand the investigation over to somebody who could fuck up a cup of coffee. Remember something else; bad scientists have families too. They have children to feed; plus they also need the practice. There is no doubt about it; we need the best bad people that we can find to get to the bottom of this.

A quickie tour of the galaxy

The Universe is, no doubt, filled with a variety of alien types with different foundations of evolution. Some are technological; some are multiformational to various degrees and many are non technological and most likely, many are a combination of several different evolutionary foundations.

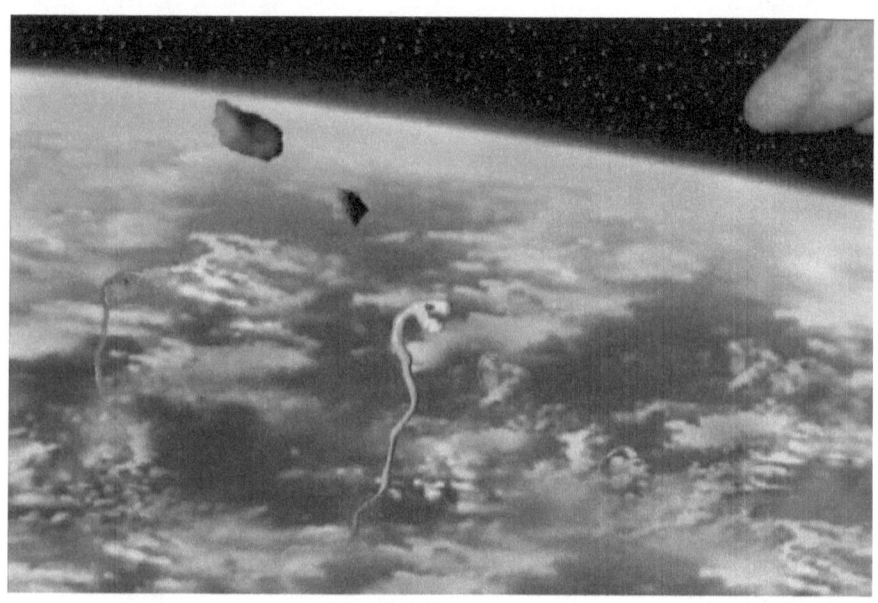

Partying with these aliens might require a ladder as they often stick their heads out of their planet into space for the purposes of waiting out violent dust storms in the planet's atmosphere.

Here we a massive consciousness consuming nutrient that lie below the surface as an astronaut to the left ponders how close he should get.

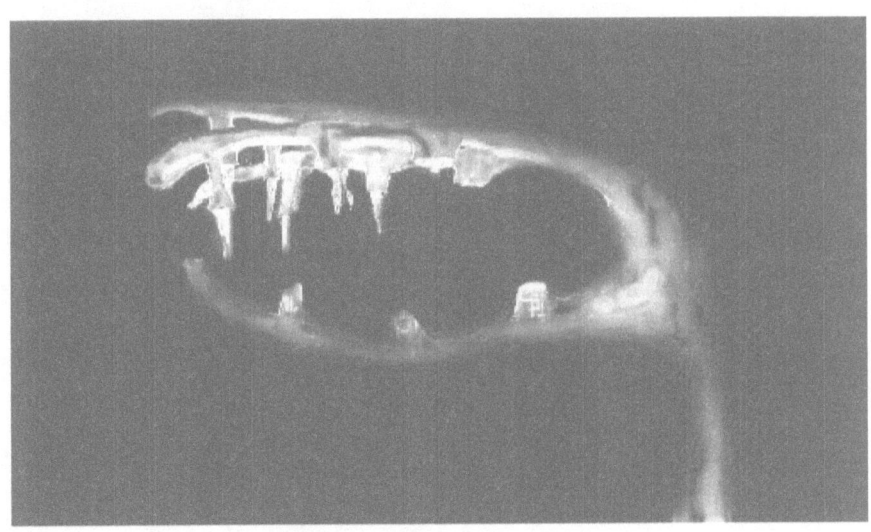

At the other end of the spectrum we see the hand of an advanced technological intelligence capable of techno dexterity that a human could only dream of.

Below we have three frames from a multitude of creatures that exist in a multiformational hive mentality. Creatures like these represent different directions of technological evolution from what a human being would experience.

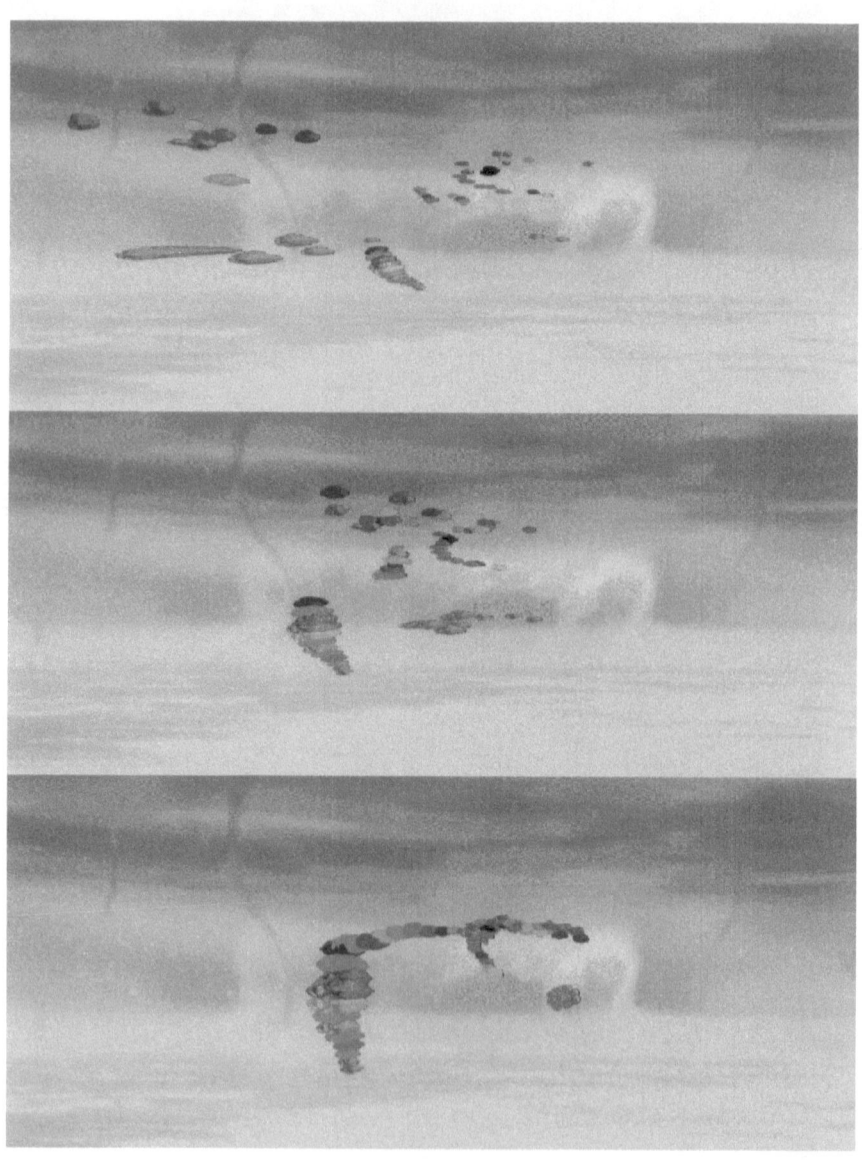

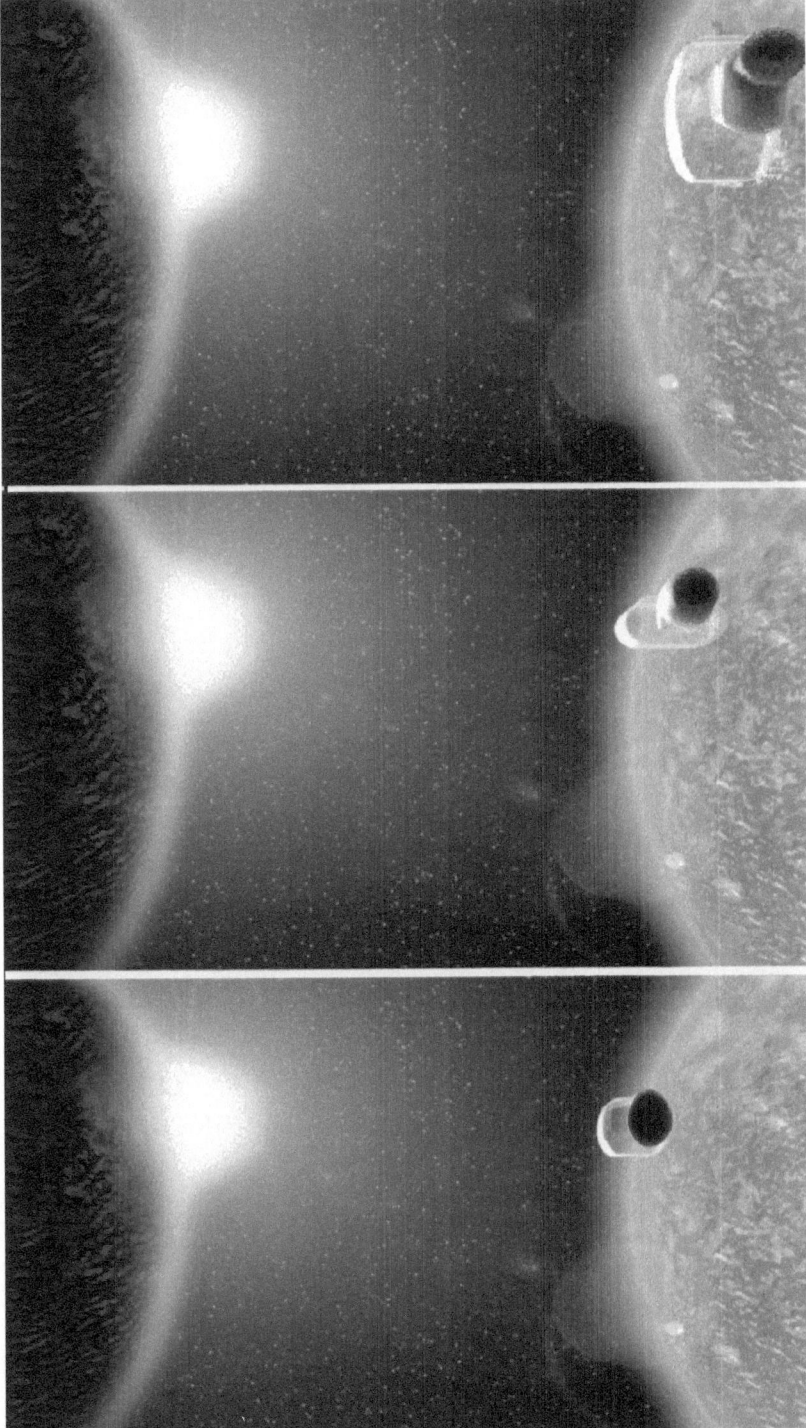

Previous: A fragment of nano genetically enhanced DNA designed by an artificial intelligence millions of years ago floats through the depths of space in search of a host of which it can assimilate.

Above; an alien of the Milky Way Galaxy; from the perspective of another alien of course; and don't kid yourself; there are aliens in the galaxy that would kill to see an image of a life form from another planet. But what would they think of it?; we see the head in the middle of what looks like a square shell it must use for protection; and the white matter that surrounds the head?; obviously these are its eggs that, of course, lie embedded within the shell for protection.

Do technological species intregate with other technological species

As has been said before most notably by the late great Dr. Carl Sagan, The chances of two alien species interacting with one another at the approximately the same time in their evolutions is remote. This as far as we can predict should only happen in Hollywood but it is likely that at some point in the evolution of a successful technological species that it would encounter some form of advanced technological species. Any initial contact between technological species most likely will begin with paranoia and suspicion. Yes, the wonderment and inspiration and awe as it normally would be described would exist but when you get down to the reality of the event of alien introduction it's always a good idea to be paranoid and suspicion, at least in the beginning ... especially the underling or lesser advance species. I know, I know, I sound like that guy in every movie where the visiting alien is a nice guy looking something really nice yet still comfortably alien that immediately declares this operation to be of military grade and eventually evolving into this person just giving the order for the aliens execution because it is important that he become more and more evil as the movie progresses on. But more than likely an advanced alien culture would not want the annulation of the human race, however, they may have their own agenda and that in itself could be a very difficult thing to comprehend. In addition it would be very difficult to even identify an agenda that might involve the human race or even the Earth for that matter let alone whether or not it is somehow detrimental. But in terms of an actual integration, this would actually begin with just the simple observation of the lower tech species of the higher tech species. The higher tech species might not be as interested in the lower tech but may theoretically be put into a position where as it must deal with the lower species and thus we may have an immediate racial component to the interaction. But the real issue of unimportance for the lower species should really be the most serious and possibly lethal aspect to be dealt with. What does a pile of ants really mean to humanity? The answer … not much! Yet ants are an evolved life form relatively speaking. Remember, we were single cell organisms for 3 billion years. The ant may very well be near the top of the tier in terms of evolved life in the Milky Way with the microbes of the galaxy occupying a much lower intelligent level yet existing in much greater numbers. But with all their achievements in evolution for more than 300 million years an encounter with humanity may produce a no respect event. It would be wise to assume not a whole lot would have necessarily changed when referring to the invitation of an advanced alien intelligence as well. Whether or not increased intelligence allows for a greater empathy of other lower intelligent beings remains to be seen. But based on humanities treatment of other intelligent beings such as whales, dolphins and elephants, it would be wise to assume the worst in the case of humanities integration in any way into an extraterrestrial matrix. Possibly worse yet might be the treatment of live 'ambassadors' by extraterrestrial gods who simply don't understand all the necessities that human beings or dolphin intelligence might require for happy healthy, productive lives. After all orcas (killer whales) arguably have brain development beyond human beings especially in the neo cortex area where the higher levels of reason and thought analysis are formulated yet we still capture and place them into work camps that are constructed in ways that are not beneficial to the orca's comfort of living. Basically, in our attempt at displaying empathy towards creatures of higher intelligence we have now demanded that captured orcas and dolphins live out their captive lives in water tanks with concrete walls that immediately reflect back the sonar that the animals use for navigation. This would be akin to a human being having bright lights constantly being emitted back at them in their navigation of their environment. Yet most people are completely unaware that this situation exists when they visit a marine park. Moreover they are completely unaware that for all practical purposes this is basically a capture and slave situation not really all that different than what African Americans underwent for more than 300 years. But it's a different perspective to view slavery and neglect from within one's own race upon one's own race. But viewing the suffering of another race of intelligence that is spectaculary different from our own race is a much longer time coming. What's worse is it take the masses to finally initiate change but to get to that point where the masses truly understand anything can be difficult to say the least.

TEX AVERY IS LIVING AT THE CENTER OF THE MILKY WAY GALAXY

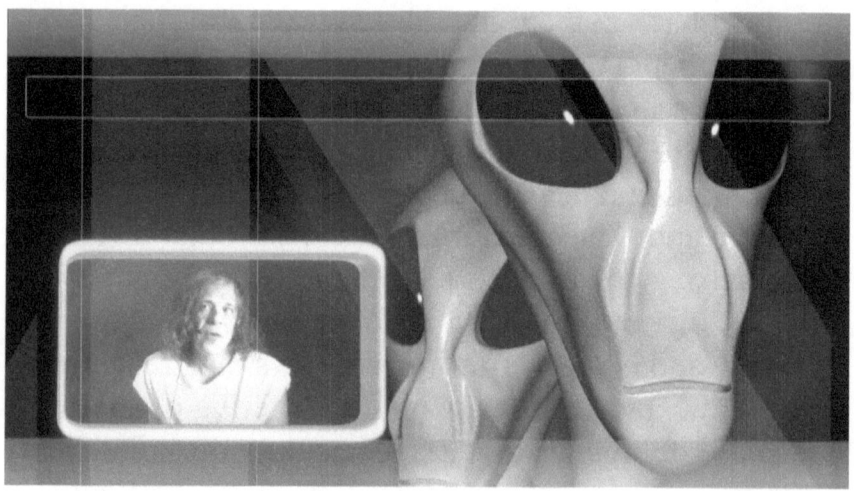

Another reason I've never been much of a gray guy is because grays don't seem to have much of a sense of humor. They always seem like they might be kind of tight ass bastards. I mean, they seem like the kind of guys that would maybe try and time a visit to the bathroom before the waitress lays down the check. Or worse yet, find some lame excuse not to tip despite the fact that waitresses depend on tips for their very livelihood. More likely, alien intelligences are likely to have a more comic or playful attitude as is the case with all known beings of significant brain development.

Everyone knows how Hollywood movies always portray alien civilizations as coming down here and committing total devastation. Trust me; we're not gonna get a plasma beam from deep space. We're going to get a pie; a huge 100,000 mile long pie.

Advanced alien intelligences are not going to be tight asses. It is a misconception which has created a perception that higher order beings stand tall and serious in their dominance of everything. Most likely they will be comically insane. The reality of it is that they are more likely to pull a fast one on you if you turn your back on them like a puppy or a dolphin or a human. But the level of super evolved shenanigans would be a frightening prospect for the unsuspecting pea brained human that may happen to wonder into the path of an alien prankster. And while it is probably impossible for us to imagine that extreme level of intelligence, some of the fun that is played out may have a common foundation to what we already understand. Imagine nano technologically advanced whoopee cushions a million years in advance of anything we ever imagined.

Above: A alien whoopee cushion waits patiently for an unsuspecting god to wonder into it's trap.

Imagine exploding cigars capable of tracking their victims across entire solar systems. These are the priorities of Gods as much as anything; wonderful healthy play. The actuality of conditions created in a super conscious intelligence network that is accelerating faster and faster towards the center of the galaxy may resemble a comic insanity network reminiscence of a Tex Avery cartoon more than anything else; total chaos. It would be constant comic creation in a rapidly evolving evolution that changes literally every second. It is beyond our present homo sapiens comprehension as to what it would be like to inhabit such an environment. In a normal Earth like setting we might evolve from the video tape to the Compact Disc in a matter of years. But in the evolution of a super conscious intelligence the entire environment may undergo massive evolutionary upgrades every second. These are not mundane, boring beings of serious dominant thought, but comic creatures who exist in a spectacular childlike consciousness. These are creatures who exist in super sexual odysseys of love and comic exchange; creatures of constant creation who literary wake up to a constant evolution of their environment happening all around them. Incredible as it may seem there may even be creatures of atomic dimensions beyond the speed of light who's environmental and perhaps even biological evolution evolves a complete metamorphic change in intervals of less than a billionth of a second. This is, of course, ridiculous to try to imagine and a million years from now it may still be ridiculous for us to try an imagine because we are now talking about evolutions billions upon trillions of years into the future. All of this probably happens because higher order mammals have to evolve in the direction of comic nuttiness as a mechanism for surviving in an evolution Universe.

Tex Avery, the great cartoon animator that deformed and perverted the laws of physics so bad that certain gods issued a warrant for his arrest, is living at the Center of the Galaxy; not these tight ass gray bastards.

Butch Cassidy and the Sundance Gray

However, one of the most fascinating cosmic

questions I've always had is what is the ability of an alien species to track us over a 100 or a 1000 or maybe even a million of years of evolution? I mean if they were aware of us 100,000 years ago; would they still be aware of us now? Don't ever think that there won't be holes in future digital records. War, fascism, and just changing interests can severely interrupt the data library of humanity and possibly anything else out there ensuring that what we might consider 'engraved in stone' information to be lost forever. An alien civilization will also, no doubt, experience the mirryadd of different ways that information can be lost or distorted in this dimension. And when we consider the spectacular distances and timeframes that must be trans versed it would be a miracle for an alien civilization to maintain a consistent observation of a more primitive species for even thousands of years. Interests change, governments change as well as the overwhelming intake of new knowledge and subjects that could easily shift the interests of even the most curious of minds.

It is also very possible that an alien civilization could visit the Earth 20,000 years ago and then another alien civilization of the same species visits again 20,000 years from now and discovers that their own ancestors were here thousands of years ago; and as a result, starts to study themselves. It's an intriguing question because we have on Earth probably the largest concentration of gold within 50 light years in both Fort Knox and underground vault in the Federal Reserve. And don't ever kid yourself; there are things in this Universe that would love to get their hands on that gold; hideous things. But you have to figure; if you have an alien civilization capable of making long distance Jumps between the stars and can also observe the Earth and are aware of approximately what evolution we're going to undertake; it might be possible for them to time a visit back to the Earth at a time when they know that we will have concentrated all that gold into one or two places and then move in for the score. More likely if greedy alien bastards tried to pull something like this off they may even try to help us out a little along the way to get the process started; maybe to speed up the process to make sure it goes their way. Maybe they might throw us a bone like in '2001: A Space Odyssey' that way we can start beating the living shit out of each other and in a million years or so, we could have hubcaps and bowling balls and, of course, have all that gold in one or two places for easy pickup.

It is, of course, unlikely the alien beings are going to come and try to rob the Earth of its gold. The fact is that they're so far away in terms of space and time that it would be difficult, to say the least, to be at the right place at the right time in the evolution of a species for maximum gold extraction. Plus, if you have a problem in deep space it's not like you can just pick up the phone and dial customer service. On the other hand if a program or artificial intelligence were planted or placed in a position to monitor the evolution of humanity for just such purposes then it might just be a bit more plausible. Still if an alien intelligence has evolved the ability to travel between the stars it is unlikely that they would have a need to bank rob a planet of an emerging technological species; especially if they knew that doing so could theoretically ignite a nuclear holocaust in that species; that; combined, with the vast distances in the space between the stars and the huge expanses of time between the evolutions of life forms make it all very unlikely. But still, it would make a hell of a movie.

Atomic Dimensional clarification

In this book I make the clarification that we are living in an atomic dimension which means it would probably be a good idea that we get a little more precise when referring to the Universe as composed just of stars and galaxies. To be precise we are an atomic dimensional matter. That is our dimension. The 'what' and the 'where' are the same thing. All the stars and galaxies and people that we have ever known are one kind of atomic dimensional matter. The whole Universe on the other hand, is everything that can ever be. This probably includes an infinite spectrum of atomic substance dimensions of which our atomic dimension of matter is but one…probably on a velocity spectrum. To be clear the Universe is everything and contains all of the atomic dimensions.

DIG THOSE CRAZY PHYSICS

Over time it may be that alien civilizations flock towards the center of the galaxy in order to fall into the higher velocities that are needed to achieve atomic dimensional transformation.

It's always been my belief that newly born technological civilizations immediately flock towards the center of the galaxy upon reaching their birth of technological construction in outer space. What is the great cosmic lure you might ask? It is the tremendous velocities that can be obtained from the massive black holes near the center of the galaxy that will allow technological civilizations to journey there for the purposes of atomic dimensional transformation. In effect they will go super light dimensional and escape into the higher velocity dimension ahead of this atomic dimension. Alien intelligences far more ancient than our own have an understanding of this; humans and dolphins will to; perhaps within 1000 years.

Wormhole 66

Ancient intelligences position themselves in orbit around a massive black hole for eventual atomic dimensional transformation. It is important to note that they do not fall into the black hole or travel through a wormhole created by the black hole; instead they use the tremendous gravitational velocities to accelerate around the black hole.

But where have all the aliens gone? If the Universe really has more stars than grains of sand on all the beaches on Earth, then where the hell are they? There has to be a reason why we have never detected them either by radio or by direct contact. The reason, I believe, may be because newly born technological species may leave this dimension through velocity almost immediately after reaching their own mass communication birth. If this is true then alien intelligences would, in fact, not inhabit the galaxy as presented in most science fiction; but instead would, most likely, leave this dimension only a cosmic second or two after their birth of technological construction in outer space that would occur at the same second as their mass communication birth; in other words; about a 1000 years. This would mean that they would most likely flock towards the center of the galaxy, as perhaps all newly born technological species do, and achieve speeds great enough to exit this dimension. What this really means is that once they reach their births of technological mass communication, they are only a few hundred years from achieving speeds beyond the speed that light travels, meaning they will leave this dimension and we will no longer be able to detect them. This may also explain why we have never detected them through radio spectrometry; because, in a general sense, once a technological species achieves space flight it will disappear from this dimension in less than a cosmic second; the reason is because they are traveling so fast that their mass has transformed into another mass and by definition, another dimension; theory only, of course; but for now let us consider that only a certain number of technological evolutions will occur, say over a period of 1 million years. Actually my best guess is that a technological civilization emerges in the Milky Way Galaxy about every 40,000 years or so. Where did I get that number from you might ask? It just popped in there one day and probably means nothing but I always liked it. The fact is however, that nobody can, even remotely, determine how many alien civilizations exist in a given area of the galaxy so everybody basically has to guess at some point. There are fairly sophisticated ways to guess such as the Drake Equation which allows educated guesses on alien variables such as how many life bearing planets exist that are capable of supporting life or how many technological species ultimately destroy themselves; but in the end one still has to guess. 1 in every 40,000 years is my best guess and probably accurate to a percentage of at least 1 in 500,000 percent. But if this were even remotely close it would mean then that at least one technological life form has evolved within a distance of only light years from the Planet Earth in the past million years. If they emerged a million years ago as a species capable of technological construction in outer space then they would have disappeared dimensionally about 1 million years ago at a time when we were upright walking apes. It makes sense then, given the vast amounts of space and time that we should not detect them; at least not yet. But also remember that we humans have only had radio telescopes for 60 years or so; a micro window of which to catch an alien transmission that just happens to be pointed in our direction and for us to receive. That doesn't mean that they are not here. It just means that the galaxy is either not brimming over with technological alien species as some have thought or they leave this atomic dimension soon after reaching mass communication birth. But how exactly does one enter another dimension?

Atomic Substance Theory

In order to understand what it means to travel super light dimensional it will be necessary to understand what the 5th dimension actually is. The following information will place the concept of the 5th dimension in your brain clear as a bell; no acid required.

Currently we have four dimensions as laid down by Albert Einstein; three spatial dimensions and one dimension of time; and now we're going to add a fifth dimension to our library of consciousness. And that 5th dimension exists in the direction of *'velocities ability to change atomic matter into other forms of atomic matter that no longer have anything in common atomically, with what that matter was before. In other words, as we see below, we could be sitting inside the interior of a star of another dimension right now but because our matter has nothing in common atomically with the stars matter then we will never be able interact with that star; which is probably a good thing.*

A good way to think of it would be;

A hole in mass is space but a hole in space is mass.

Let me put it another way; Say we have an asteroid made of ordinary matter consisting of protons, neutrons and electrons that is traveling at a normal speed of a hundred miles per second. That might sound fast but it's not really. Then by chance it approaches a black hole of which begins to then accelerate said asteroid tremendously. The spectacular velocities of the black hole will now transform the atomic properties of the asteroid to the point where it no longer has anything in common with what it was before it began its acceleration into the black hole. When it is accelerated to the point where it no longer has anything in common with what it was before then it will then be able to occupy the same space and time as its fellow asteroids it left behind; yet we will never know that it is there because there will be no reaction because they have nothing in *common in terms of atomic substance. In other words it will be dimensionalized. You will not only be someplace else but you will BE something else. When it comes to comprehending the 5th dimension of atomic substance the 'what' and the 'where' are the same thing.*

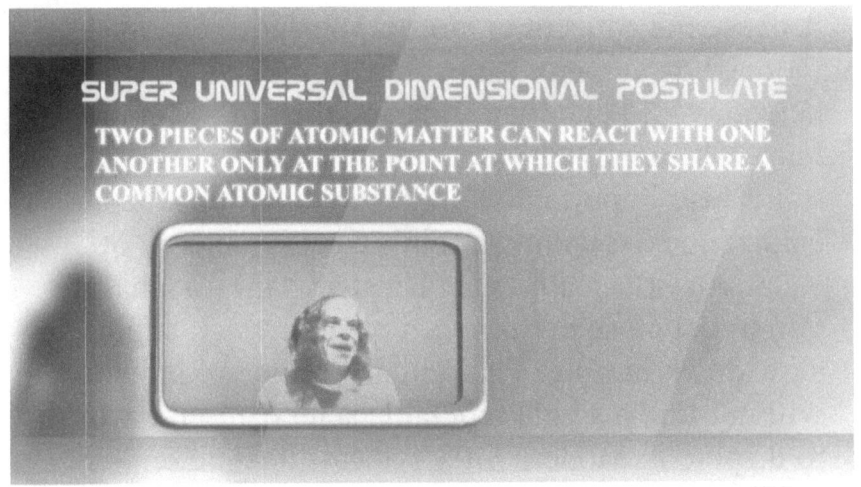

SUPER UNIVERSAL DIMENSIONAL POSTULATE

TWO PIECES OF ATOMIC MATTER CAN REACT WITH ONE ANOTHER ONLY AT THE POINT AT WHICH THEY SHARE A COMMON ATOMIC SUBSTANCE

Understanding atomic dimensions means imagining one kind of atomic substance having nothing in common atomically with another atomic substance.

It will take you from 72 to 700 hours to fully comprehend three dimensionally the fifth dimension inside your brain; but the 5th dimension of atomic substance change can be conceived in the human brain clearly; And yes, you can do it three dimensionally. In fact, you can even illustrate it two dimensionally if you want to; and it is important to have the concept presented above in your head as a visual transformation as opposed to just being able to state it logically using the English language. It not like a tesseract or superstring dimensions where a particular way of looking at it only suggests that another dimension might exist. Velocities ability to transform matter into other kinds of matter is a concept that can be clearly perceived in the human brain without the use of the English language. The reason that it will take from 72 to 700 hours is because the concept of atomic substance change as induced by velocity is not something that a normal person would experience during the normal course of human lifetime.

It is also important to understand that this theory cannot coexist with Einstein's Special Theory of Relativity. Basically what Einstein said was that if you travel close to the speed of light you'll begin to slow down in time. This actually happens and you will not find a single reputable scientist in the world to dispute it. Modern GPS systems depend on employing relativistic equations into their calculations otherwise your GPS would be off by around three hundred feet. After Einstein conceived his theory of relativity he then went on to understand it as mass and gravity actually bending space and time. But this was a misinterpretation of what was actually happening. Mass and gravity do not bend space and time. What you're seeing here is velocities effect on mass. It's all about what happens to matter; not to space and time. Space and time are static infinite. They cannot be manipulated but are ideally suited to allow matter to get into whatever trouble that matter can get into. And matter can get into a lot of trouble; remember the extinction of the dinosaurs? Remember nipplegate?

This notion that mass and gravity actually bend the very fabric of space and time maybe one of the biggest misconceptions in the history of science. I mean people have been born, had careers and then died believing this. It's on par with the Flat Earth Society. We are living in flat Earth times right now.

What a Spectrum of Atomic Substance means in the Universe

It is important to understand that Atomic Substance theory and Einstein's theory that mass and gravity actually bend the fabric of space and time are not compatible. In other words only one can exist. The world has a choice; it can believe the ideas of Albert Einstein; winner of the Nobel Peace Prize and arguably the most respected scientist in all of human evolution or they can believe me, Paul Vancleve who spent most of his pathetic life working at the local Walmart barely making ends meet. Oh, by the way, I was never promoted in nine years. It's a tough choice; I know. Keep in mind though that only the notion that mass and gravity bending space and time are being challenged here. If this notion is replaced by the idea that velocity actually changes mass then it will rewrite the structure of the Universe yet all of Einstein's equations will still remain intact. You have to understand that in order for this theory of velocity changing matter into other types of matter out of this dimension to be a reality then I need all of space and time un manipulated. I can't have Einstein's mass and gravity bending the very fabric of space and time because in my theory I need to have a spectrum of atomic substance spanning infinity over every given point in space and time. If I let Einstein just distort the hell out of space and time with mass and gravity then it would have to manipulate space and time for the entire spectrum of atomic substance generated throughout infinity. I need all of that space and time for myself. I'm a selfish bastard, I know; but one of us has to go. He can't just spread his 'mass gravity distortion of space and time thing' all throughout the infinity of space and time; that's private property. I don't care if he's dead or not.

In short; with Einstein it was all about the manipulation of space and time; with atomic substance theory it is all about the manipulation of mass.

But Atomic Substance Theory says far more than what happens to matter that is accelerated to beyond light speeds. No longer do we need to look at the Big Bang as the explosion that created the Universe. We may now view the Universe as an infinite funnelization in and funnelization out of atomic matter through velocity dimensions. This would mean that matter would be funneled in from the slower velocity dimension below us through quasars and then funneled out of this dimension through black holes into a faster velocity dimension ahead of us.

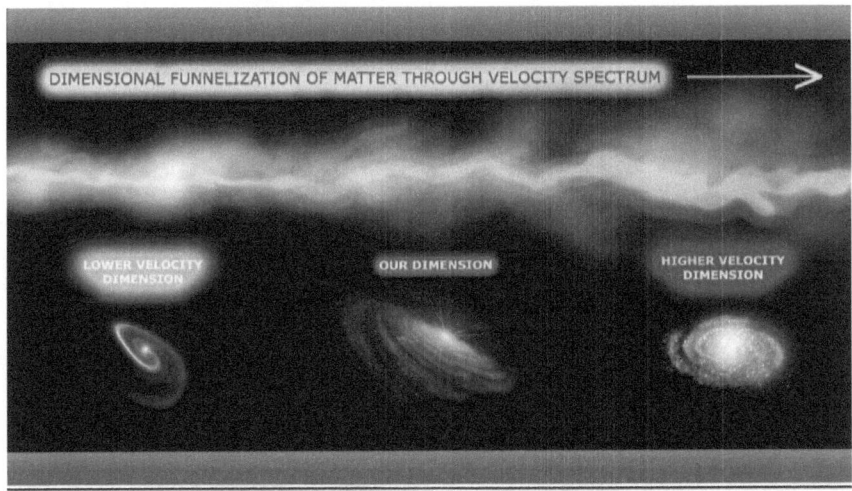

This would also suggest that all atomic dimensions have a slower velocity dimension before them and a faster velocity dimension ahead of them; like any random number has a smaller number before it and then a larger number after it. In other words matter is constantly being funneled into this dimension from perhaps quasars and matter is constantly being funneled out of this dimension through black holes. This is something that advanced alien intelligences would understand and employ shortly after reaching their own mass communication birth. And they, no doubt, have developed and immersed themselves into a dimensional astronomy utterly crucial in their journeys inward towards the Center of Galaxy. Once they reach the outer gravity pull of the black hole they then begin to travel around the hole at a precise arc that will increase their velocity ever so more allowing the velocity induced atomic substance change of their mass to form just ahead of the previous velocity layer of atomic substance mass that they would have crashed into. This would be like falling out of a building and just before you hit the concrete the entire ground begins to fall just ahead of you thus allowing you to continue falling indefinitely. The reality of this is that the changes in velocity induced atomic substance happen at a rate that allows the falling ship to not impact the black hole matter if it falls at a precise arc and velocity into the hole. It's a little like when a ship reenters Earth's atmosphere at a precise angle to survive the heat except now we are entering the black hole at a velocity arc that will allow the ship to transform its matter into another kind of matter at a rate that it never impacts the ever deepening black hole matter that is always ahead of it. Yes it's difficult to understand but the diagram on the next page may help.

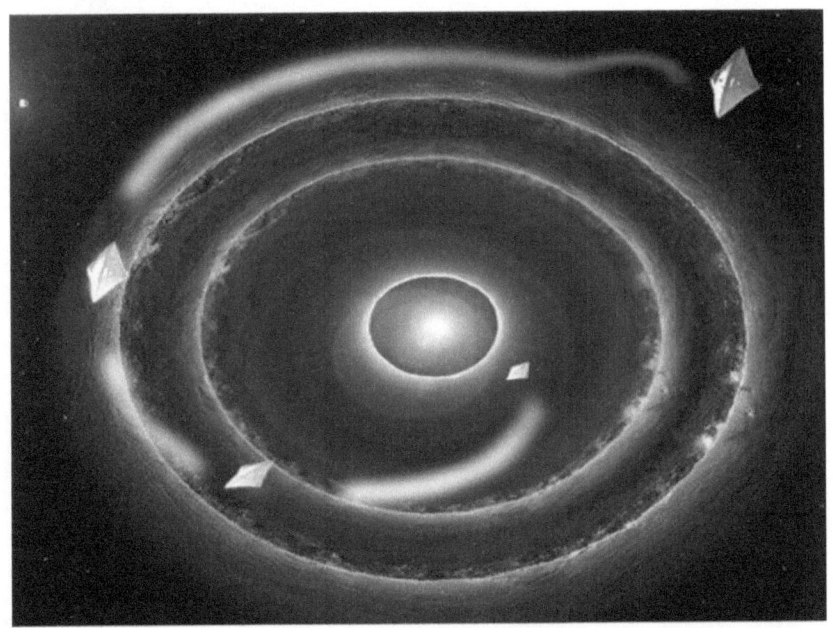

The velocities increase ever so more as we approach the center of the black hole. Our starship enters the black hole from the top right and in an arc gaining in velocity as it curves around it. The increased velocity transforms the ships mass into the atomic substance of the inner darkened circular region thus preventing it from crashing into the outer fire ring near the event horizon. That's because it no longer has anything in common atomically with the outer fire ring.

Then the same thing happens again. It transforms into a new velocity matter that is the same atomic matter as the green ring thus preventing it from impacting the mass of the darkened ring of matter that it appears to be in physically.

Then when it reaches the point near the center, the ever increased velocity has transformed the ships matter into the same matter as the inner yellow ring thus preventing it from crashing into the green ring of atomic substance that it is flying through. Got it! The whole thing should work great except for the Spaghettification that would rip your guts out in less than a micro second.

Previous: Three frames of a hypothetical atom in motion in another dimension. The sphere would be similar to an electron and the rectangle would be similar to a proton. The sphere bounces off the spinning rectangle in a balance of attraction and repulsion. However in this atomic dimension there is no fusion to propel matter back out into space for the formation of life, so all of the matter just collapses in on itself into an infinity of black holes forever.

Dark energy must die because it doesn't exist

One of the fortunate casualties of Atomic Substance theory is the death of dark energy. Forgive me if I sound a bit like Lex Luthor here but I've waited a long time for this day. And yes, I do get a sick thrill out of watching dark energy die…slowly. Die you galaxy sucking bastard that doesn't actually exist! Ok, that was unnecessary, I admit. But let us remember what dark energy is … if it actually existed.

A long time ago the great Edwin Hubble discovered, using the Hooker telescope at the Mount Wilson Observatory that the galaxies that were furthest away from the Earth seemed to be accelerating away from us at a faster rate than the galaxies that were closer to us. In other words they were red shifted to a much greater degree. This was a true mystery. Was there something special about the Earth? Did we smell or something? Nobody knew; so because there was no obvious answer to this anomaly then the idea of 'dark energy' was invented, some years later, to explain why it was that the outer galaxies were flying away from us at much faster rates. It didn't involve anything about the Earth as being unique or anything like that; and as it turned out, no matter where you were in this atomic dimension you would always see the outer galaxies expanding away from us. A good way to think of it is by imagining a balloon expanding as it fills with air; no matter what point you are at on the surface all of the other points will be moving away from you; except in this case, the further the galaxy was away from us the faster its acceleration away from us. The idea was, of course, that dark energy was pushing or pulling the outer galaxies away from us. It was an odd theory that never felt quite right; at least to me; and it really is living proof that even in the scientific world of high level interpretation, if the answer doesn't reveal itself after a certain amount of time then an answer will be invented. It was also, strictly, an inferred theory; although it should be noted that some scientists actually think that they may have detected it directly. Dark energy was one of those predicaments where they had to say something, I guess; and its imagined existence may very well be the result of thinking of the Universe in finite terms such a age. And while most scientists are open to the idea of the Universe being infinite; interpretation of the Big Bang Theory has unwittingly directed thought into finite terms; but understanding the outward acceleration of the galaxies may require an answer involving infinity. Atomic Substance Theory allows for the infinite funnelization in and the infinite funnelization out of atomic matter. Which suggests the viable theory that an infinite amount of matter exists outward into infinite space that should a have a gravitational effect upon our observable 13.6 billion light year radius sphere.

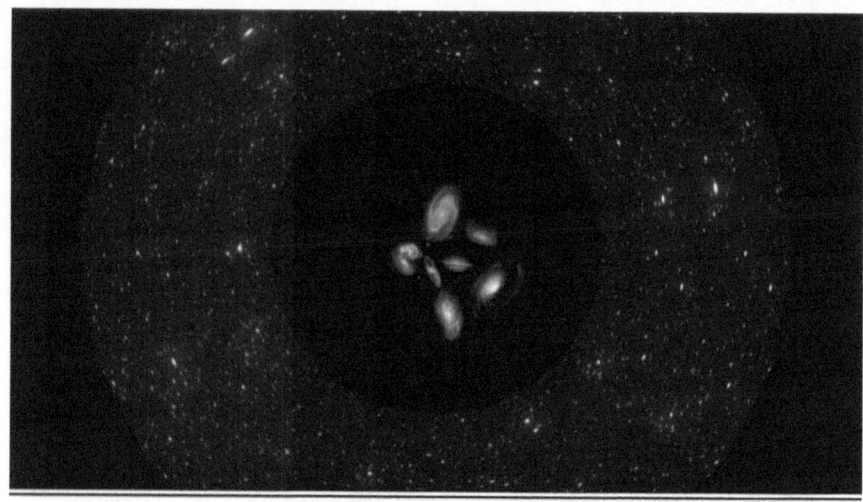

The galaxies in the middle represent the observable universe and the outer ring of stars represents hypothetical matter that extends outward into infinity. If we think of the universe as infinite then this infinity sphere of matter should exist and also have a gravitational pull on the outer observable galaxies that are accelerating away from us.

Because Atomic Substance theory suggests that matter may be being funneled into this dimension from another dimension then we are no longer constrained by the idea that the Universe is 13.6 billion years old…or 92 billion years old as speculated by some who have attempted to calculate the movement of all known mass from its original space of origin. In other words we are now able to view the Universe as, perhaps, being infinitely old. And because the Universe may be infinitely old then we could safely assume that there is ordinary baryonic matter no matter how far out you travel. This means that if you travel 13.6 trillion light years (yes, with a 'T') out into space then you will still find ordinary matter in the same various states that we find it here; which means that there must be an infinite sphere of interstellar matter surrounding our little 13.6 billion light year radius sphere of observable matter; which means that this infinite sphere of matter must be gravitationally pulling outward on our observable sphere; specifically the outer galaxies that are, in fact, known to being accelerating away from us; which means, of course, that it is the infinity sphere of matter, that exists into the forever of space, that is causing the gravitational accelerations of the outer most galaxies and not the mysterious dark energy that ejected from someone's butt out of deparation to find and answer. But there could also be another answer …

The Outer Colossal Black Hole

This would be where 99.9999999% of all matter spends 99.9999999% of its time. Feel free to add a few 9's onto the end of that decimal if you wish. These insanely huge black holes that never formed from any stellar evolution, but instead from the infinite debris of used matter rubbish, may be what is pulling on the outer galaxies to cause their acceleration. Theoretically, galaxies that eventually make their way beyond the point of knowing of the existence of other galaxies, or for that matter, have themselves, lost the ability to produce stars, would accumulate somewhere in the distant space and time into massive black holes beyond anything that ever formed in a normal galaxy. These outer colossal black holes could, if they actually exist, be as large as the current visible Atomic dimension or visible Universe. And if they do exist, then they probably give off energy in the form of a deep dimensional fusion. Lets face it, the speeds and velocities at the center of something like this would be beyond anything imagined and probably would transform matter into other kinds of atomic matter for dimensional transformation.

We see a general synopsis of the Big Bang explosion moving from left to right. At the far right in blue we see the mysterious Dark Energy accelerating the outer galaxies away from us.

.

In this above image from left to right we see matter being funneled into our dimension through quasars, which may be white holes that exist on the other side of black holes in the slower atomic dimension. As we move to the right we see the outer galaxies being accelerated, not from Dark Energy, but from the infinity sphere of matter.

Time travel without pissing off any Gods

Time itself, which is the propagation of the physics of atomic matter in the direction of cause and effect may actually curve around itself like the galaxies making relative time travel possible. The key word being 'relative' meaning that there may be a dimensional aspect to time travel that prevents any paradox from occurring.

Another aspect that Atomic Substance Theory suggests is the possibility of time travel WITHOUT PARADOX!. I say 'without paradox' because time travel into the past has always had an immaturity or implausibility aspect to it; at least when it comes to actual science. Most of us have heard of the classic time paradox where a person travels back in time and kills his own grandfather thus preventing his own birth years later; and this has always been the classic 'deal killer' for any substantive consideration of actual time travel. Time travel in the movies is great but if something like 'Back to the Future' happened in the real Universe then it would bring about the non existence of the entire human race as well as all known ecology past and present. This is because eventually if the common man had the ability to travel back in time and actually change events then eventually some time traveling terrorist penis would travel back far enough to stop all life from forming on the Planet Earth; and why would he do this, you might ask; could be a lot of reasons; to stop his girlfriend from sleeping with his best friend; somebody else got the that big promotion; who knows? But fortunately the real possibilities of time travel into the past will most certainly involve traveling back in time only RELATIVE to another time period. The reason it may be possible to travel backwards in time is suggested by the fact that time dilation causes a spaceship that is traveling at or near the speed of light to slow down in time relative to the Earth. This, at least, suggests that time travel into the past may exist in the direction of increased velocity because normal time is slowing down for a spaceship traveling at relativistic speeds. Also it may be that time, like the galaxies, may be spiral in shape? But in reality, it would more accurate to suggest that matter, as running through the spectrum of atomic substance, may be curved relative to itself. This suggests that maybe velocity is slowing down time for matter to eventually, not just slow down in time, but perhaps begin to travel backwards in time relative to the Earth.

In addition, as velocity transforms matter into other forms of exotic matter beyond the speed of light, then it may very well be that we find that the spectrum of atomic substance, as generated by velocity, and as a whole, is curved like the galaxies. And it may very be that the crushing temperatures at the centers of black holes create those speeds that allow new kinds of matter to be created to become the faster dimension ahead of us and allow travel into the past. Take a look at the filmstrip below;

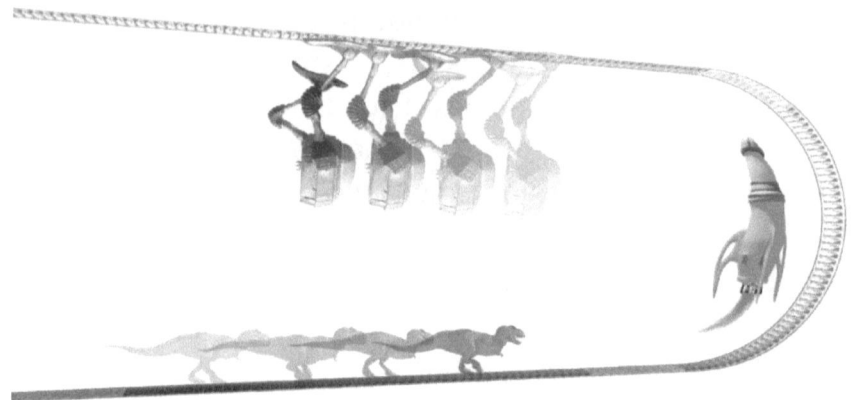

The tyrannosaur sees time as moving forward normally. Then we see atomic matter curve around itself with increased velocity beyond the speed of light. Then we see the robot being above walking forward and also experiencing time normally; but because time is curved around itself the tyrannosaur and the robot see each other moving backwards. More importantly, the Robot is traveling backwards in time relative to the tyrannosaur. Remember also, the robot is traveling faster than the speed of light so his atomic matter no longer has anything in common with the tyrannosaur's atomic matter therefore he cannot affect the tyrannosaur in any way thus eliminating the possibility of paradox.

As we can see from the strip of film above everybody always see time as moving forward. In other words the toothpaste will always appear to come out of the toothpaste container exactly the way it does in the commercials. But as we can see, the strip of film, which represents the flow of time or more correctly put; the propagation of physics (cause and effect), is curved. Ultimately this curve is made up of pure velocity as a foundation. The t-rex at the bottom sees time flowing normally and the robot at the top sees time flowing normally. But relative to each other they see each other as moving backwards. The t-rex on the bottom sees the robot on the top moving backwards and the robot on the top sees the t-rex on the bottom moving backwards in time relative to himself as well. Now remember this filmstrip not only represents the flow of time but it also represents the velocity of mass. The t-rex on the bottom is in our dimension but the ship in the middle is accelerating to super light dimensional speeds (faster than the speed of light). The robot on the top is traveling well in excess of the speed of light; and because the robot on the top is traveling in excess of the speed of light velocity has transformed his matter into another kind of matter that no longer shares any common atomic properties with ordinary matter that we all know so well. So there is a dimensional aspect to time travel here; and this is what eliminates the old 'kill your grandfather and prevent your own birth' paradox. The robot on the top is now traveling back in time relative to the t-rex on the bottom to the point where they are both at the same time period relative to the Earth. But because velocity has transformed the robot matter into something else then he can no longer interact with the t-rex on the bottom; meaning NO PARADOX! But there is nothing at least in the way of creating a paradox that prevents him from seeing the t-rex on the bottom. By the way I have no idea in hell how he would see him but the point is that there would be time travel without paradox.

But what if an astronaut could simply see the past? What if an astronaut arcing his way around the great spiral arm of time and into the past, relative to when he left, could somehow see into his dimensions past history and record the events down to an atomic particle level. What if he could actually see the colors of the strips on the underwear of people on the Titanic? How kinky would that be? He could then answer any question ever poised in history. If he wanted to know who killed Kennedy he could know and record every micro aspect of it. The movements of every atom could be recorded and perhaps interpreted in terms of evolution. He could see into the brains of anyone and know exactly what they were experiencing. Eventually our astronaut could even map out the brains of anyone he wanted and perhaps upload copies of their minds and consciousness into his new dimension into the past which lies in our future; confused yet?. He would actually do this by reconstructing their brains atom per atom into his new dimension. It wouldn't be like cloning where we just get the identical body and not the consciousness. By being able to reconstruct a person atom per atom we could actually replicate a person's exact brain consciousness and personality to be alive somewhere else; presumably beyond the speed of light. Perhaps Kennedy or Lincoln, or maybe even Einstein have been recreated atom per atom into the past and now reside beyond the speed of light engaging in God knows what kind of cosmic experience.

A human astronaut of the deep future that is traveling faster than the speed of light travels back in time to view the sinking of the titanic. There is dimensional separation here so the astronaut can never actually affect the sinking of the ship.

As I said before velocity can transform matter into matter that no longer has anything in common with what it was before. This means that matter that has been transformed by velocity beyond the speed of light can now occupy the same space and time with ordinary matter as we know it and neither will ever know that the other is there. It is the type of mass that is the dimension, not the space. A good way to remember it is; a hole in mass is space but a hole in space is mass. This means then that there is a dimensional aspect to the possibilities time travel which means that if I traveled back in time to the Kennedy assassination that I would have to become another mass of another dimension to do so; which means that I could be in the same time and space of the actual assassination but because my matter has nothing in common with ordinary matter there could be no reaction. I might be able to see it somehow but I would be prevented from actually interacting with it. This means that no being or God no matter how powerful can ever affect the past. The past is forever safe from manipulation. It also means that the future of humanity in space may take place in the past. If it is true then right now our descendent star children relatives could be watching us and our every move. Yes even in the shower!

However, something incredible may be happening to us in the past as we speak. Super documentation of the evolution of the Earth from its earlier primordial fetal development to its spectacular evolution of conscious 4 1/2 billion years later could conceivably have been achieved by intelligent beings billions of years beyond our primitive detections. Sometime, a million years from now, we may achieve access to Earths primordial evolution 5 billion years before our existence. Perhaps even as we speak, super images of the planets and stars of this dimension are being funneled in towards some huge outer universal depository for record and analysis in keep, The Super gods, now a consciousness spread into vast arrays of technological and biological conscious integrations will now record or attempt to record the infinite past in all respects right down to the level of the electron. This means that the previous movements of every atom that ever existed in the past will someday be tracked and recorded for all of tomorrow to view. As a matter of fact, it may very well be that the movements of every atom that have ever occurred in the history of the Earth could, in fact, be a matter of record. You could actually look it up if you wanted to. All you would need would be a library card and a starship.

And, of course, the opposite may be true if we could somehow travel slower than this dimension. I'm not exactly sure how a person would do this; even in concept; but I guess the idea of accelerating against the galaxies would be a start, conceptually at least. But think about it. If Einstein suggested that we may travel back in time by traveling at or beyond the speed of light then logic suggests that traveling slower than this dimension may land us in the future. That's weird! But that's the logic of it. So it may be that mass that is funneling in from the slower dimension below us, which would then be from the future, through our dimension, and then out of this dimension into the faster dimension ahead of us though black holes; suggests that maybe our atomic dimension of initial plasma/hydrogen may have exploded from the future; which means that our evolution in to humanity may ultimately have exploded from the future. There may very well be beings billions of years beyond our evolution who, in fact, have traveled backwards into this slower dimension and perhaps forward into time to the origins of where the matter in this dimension was ultimately funneled from.

What if the speed of light and 186,282 miles per second are, in fact, not the same thing.

I came up with this theory a few decades ago and while I'm not quite on board with it anymore; it deserves special mention. According to Einstein's special theory of relativity, nothing can travel faster than the speed of light; and the speed of light is 186,282 miles per second. What Einstein is really saying in his equations is nothing can travel faster than 186,282 miles per second. The reason that it is not possible to travel faster than the speed of light with any known energy conversion process in this dimension is because the amount of mass increase in matter that is being accelerated is always a step ahead of whatever amount of energy that you use to accelerate it. This is why occasionally you will hear a scientist say that you could put all the energy of the Universe into your rocket and you still wouldn't be able to surpass the speed of light. The reason is because the more energy you put into acceleration then the more mass you gain which means you need more energy to accelerate the new mass which causes an increase in mass which means we need more energy. It's a catch 22. Think of it like this; if we both walk in perfect step unison for a billion years but I start out one step ahead of you then guess what?; after a billion years I will be one step ahead of you.

The speed of light has also been clocked accurately at 186,282 miles per second. It sounds like they are the same thing but that may be because it was only **assumed** they were the same thing. What if the speed of light and 186,282 miles per second are not the same thing. What if for over one hundred years we only assumed they were the same thing because they were so close to one another. What if there is a billionth of a degree or so difference (not the exact number) between the two? And as long as we don't surpass 186,282 miles per second, we are in accordance with natures laws; and don't worry, we won't. What would this mean if it were true? It would mean that we are free to forever mount the decimal points closer to 186,282 miles per second forever into infinity. And what, may you ask, exists in this last billionth of a percent of this dimension between the speed of which light travels and 186,282 miles per second; an infinite spectrum of atomic substance expanding into forever; an infinity of matter spanning into an infinite spectrum of atomic substance fractaling forever into the infinite breakdown between the speed that light travels and 186,282 miles per second. It would be an infinity of dimensions perhaps expanding backward in time into the past.

Other ways to 86 this dimension

There are other ways out of this atomic dimension so don't get depressed. And they all have the common denominator of atomic substance difference that makes them exist as separate dimensions from one another. The one discussed above, of course, is velocity and its transformation of mass. But you can also go the chain funnalization way as well; an even more kinky direction. In this direction atomic substance A is connected to atomic substance B which is connected to atomic substance C which is connected to atomic substance D. and so on. Atomic substance A can only interact with atomic substance D through B and C. A hypothetical example of this might be if we imagine a Proton and then imagine a type of quark that is connected only to a proton. We then must imagine another type of quark that is connected only to that first quark and then on and on; you get the idea. Its like a chain with each link connecting only to the link next to it.

And yet there is even another way out of this universe; in fact, it is two ways out. It is the old 'our universe is only a particle sitting in a plate of spaghetti theory' that I'm sure you've heard of. And remember, it has to be spaghetti; fettuccini doesn't work. This is kind of like a 'Horton Hears A Who' concept where as we expand out to imagine our dimension or Universe, which ever you prefer, as being a super micro small particle that is really just another atomic particle that is part of something much larger of which your imagination if free to speculate. It's important to understand the way this theory works in terms of atomic dimensions. We have gravity; if you were to look at this dimension from the distance of it being only a tiny speck it might be that all of gravity comes together to form a new force which then by definition would make it another dimension because as a whole it is a different kind of matter. But this way out of this dimension is just by simply getting larger and larger where many or all of the forces that we know of are just parts of much larger forces that are much larger than even the observable universe. In the same way, of course, we can imagine it the other way where as a single electron broken down into a fraction of 1/000000000000000000000000000000000000000000 00 then contains another whole dimension of where stars and planets form and then, of course, you can break it down from there in the same way even more on and on into infinity. Whether or not quarks actually break down into infinity is not known but if it were found that they could be broken down then it could probably be assumed that that Universe expanded the other way as well into the larger and larger into infinity as well. If life actually exists in the larger and smaller directions of infinity then it would be logical to assume that in the smaller direction of quark breakdown that whole solar systems would form and die in a billionth of a second; and that a solar system that existed in the larger 'spaghetti' direction would appear to be immobile relative to us for trillions of years. Still it doesn't mean that we can't try to be friends with them anyway.

You can't get there from here or
Are there atomic dimensions where life cannot
form and matter just loiters around for all infinity.

I always loved the phrase: You can't get there from here. It's such cosmic enigma question. Are there types of existence that are neither matter nor life. In this atomic dimension we have fusion. Meaning the fusion of stars throws the atomic matter of this atomic dimension back out into space for the formation whole solar systems that contain life. If it weren't for fusion then basically all of the matter in this atomic dimension would just collapse in on itself into forever collapsing black holes into infinity. In other words, there could be other atomic dimensions of matter where matter just floats around doing nothing throughout the infinity of forever? It would mean huge clumps of matter just floating around in darkness forever because the particle dynamics that make it up are not capable of occurring in consciousness and because we would have to have an atomic connection with this hypothetical matter we may never be able to detect their actual existence.

The effects of Atomic Substance Theory on the human brain

Something strange has happened to my brain ever since I've gained a clear view in my head of the 5th dimension of atomic substance as induced by velocity. The concept of feeling weight has become a cosmically pleasurable experience for me. But why shouldn't it?; atomic substance theory is all about speed and velocity; and speed and velocity are directly relatable to weight and force; and when you think about it, the perception of weight, itself, is yet another sense beyond the six known senses to humanity. It doesn't involve sight or sound or taste or smell. If anything, it involves 'touch' to sense it but 'touch' doesn't really encompass the entire magnitude of the feeling of weight or force. Yet weight, which is a direct result of velocity, which in itself, is the pathway out of this atomic dimension, requires more than just touch to experience. It is, in fact, the seventh sense; and with a proper and clear knowledge of what atomic substance theory is, the feeling of weight seems to allow me to feel this dimension in a way that is unbelievable. It is almost as though there is some kind of quantum brain effect that is happening. Perhaps this is because weight, force, speed and velocity exist at the core of the existence of the material universe; and by having a solid knowledge of atomic substance dimensions in association with the feeling of weight and force I feel as though I'm in touch with the Universe cognitively in a way that has created the euphoria of a cosmic awe experience.

Google Milky Way

Advanced alien life may very well be integrated into an ultra complex artificial intelligence network where massive amounts of data are recorded of which most of it will never be viewed by a biological alien.

We have explored many of the ways out of this dimension and pondered some of the ways that super advanced aliens may be processing information. And considering that so much time has elapsed in terms of the evolution of the galaxy I wouldn't be all that shocked to find that we were being watched or more likely being processed in terms of data for anybody who happens to be a god and also gives a dam. Remember the gods have lives too and can't be interrupted for every little mass communication birth that takes place. But theoretically enough time has passed that spectacular evolutions of intelligence should have taken place. At this point, somebody should be aware of us.

And so we have it that a funny thing has happened in the last month (6/19) or so; and that is that UFO encounters have taken on a new reality. It's almost as though they have suddenly become an issue for everybody including Congress. Multiple witness and collaboration by navy fighter Pilots and radar tracking instances seemed to have been so frequent that the entire UFO phenomenon has now been accepted by a much larger consensus of the global community. So we have to ask: Are there really giant 500 foot Tic Tacs zooming around the pacific? In reality, the whole Tic Tac thing makes sense. No visible aerodynamic engineering; no problem; no infrared heat signature; no problem; the ability to just shoot off instantly to 30 times the speed of sound and bank instantly at a right angle; no problem! This isn't engineering on any aerodynamic level that we know of. This is nano technological engineering and construction that is 10's or maybe even 100's of thousands of years beyond our understanding. It will be multiformational, possibly to an unlimited degree. It will have, perhaps, a million engines contained within the space of golf ball. It may depend on water to a large degree for functionality. There will be no biological aliens on board. There may not even be biological aliens who are aware of it. They could find and access it, though, if they needed it. It may be capable of, perhaps, trillions of analytical interpretations of the Universe every second. Because of this, It may have strange problems: Because it exists in a network of differentiated consciousness, It may have runaway A.I. syndrome, This will occur when the A.I. digitally simulated forecasts of future evolutions becomes too far advanced for the alien biological life forms to adapt too, meaning that the biological descendants cannot even comprehend or engage in the evolutionary advancements for millions of years, of which the A.I. intelligence has already begun to create and inhabit for itself. In a single thought and emotion, an alien intelligences digital workhorse could create technology of which the overall concept will persist for a million years before someone accesses it. It really adds up to being super, hyper intelligent wind. And, of course, it will process literally mountains of information of kinds never imagined. And, at least for me, the big question is; is there a dimensional aspect to it? In other words; is it constructed out of ordinary matter or are the physics involved extending or emanating from another atomic dimension? But here's what you have to understand: Nobody gives a shit. I'm not talking about us. I'm talking about them. Let's look at it like this: We all know what Google Earth is; right? Now what does Google Earth do?; they harness pictures of the Earth and assemble them to comprehend the Earth in a multitude of different ways, of which there is a good chance that nobody will ever examine much of the data. There could be a shitload of Bigfoots down there doing the wild thing right there on satellite imagery and nobody would ever know. Yet we have a record of it created from our own technological creations. And I have no doubt that Google Earth has already started using self driving cars with mounted cameras to map out every last inch of detail about this planet containing information no one will ever see. But you get the idea; right? What we have here is a benign super being version of Google Earth. But more importantly, this is what should be here! Should we really be all that shocked that a presence beyond our ability to imagine would be curious about us at our birth of mass communication technology?; which, I might add, is occurring in the same second as our birth of technological construction in outer space, along with the threat of global thermonuclear war, and a global technological mass extinction, all in terms of evolution?. No; lets face it; it's a hot evolutionary second! Who wouldn't want to watch? Oh yeah; don't forget about the porn! But who, exactly is who? It is a common stereotype to peg extraterrestrial encounters on caricatures of human fears and even desires. We call them grays. And what they really are is nothing more than a simplified version of ourselves in alien form imagined in the Human global brain; big eyes, big head and big hands. Does Bugs Bunny ring a bell? The reality of it is, that an extraterrestrial presence beyond our evolution will no doubt be a vast network of multiple super brained species existence. We on the Earth already have three intelligences destined to merge into conscious evolution; dolphin, human and A.I., with, of course, the dolphin at a slightly higher level of brain complexity then the rest. But a presence millions of years old could be made up of a vast network of differentiated consciousness; and the process of information acquirement will be spectacularly automated; which means, nobody gives a shit. On the other hand they could be making bets on whether or not we survive thermonuclear destruction. At any rate, its not surprising that there may well be somebody watching us. There are billions of years of life evolution that have occurred before us. So smile and pucker up. You're on cosmic camera; and it's the perfect time. Also remember something else too; an extraterrestrial intelligence is an extraterrestrial intelligence; meaning that they can't think like us, even if they are smarter. So I have to wonder about those big ass Tic Tacs. Could they be some alien god's idea of deception; maybe a god who is not the brightest bulb on the tree, so to speak, just trying to blend in as best he knows how, maybe? Or maybe they think they are just being clever. Remember, an extraterrestrial consciousness may see a giant Tic Tac shape as a way to hide from those magnificent men in their flying machines because in its way of thinking, it resembles a giant cloud. If so, they may need to work on their artistry. One other odd thing about this: Google Earth, is, I believe, taking over the human race here on Earth. I'm talking about our Google Earth on Earth. Google Earth's goal is to record data about the Earth in as vivid and communicative way as possible. And soon people will have their own avatars walking around the Cyber Earth of the future and they will be collecting information not yet imagined to exist in ways we can't imagine. Google Earth will then engulf the entire Earth and a new Earth will form in whole; a cyber earth. And that cyber earth will forever collect and store even more data for interaction with dolphin, human, and A.I. intelligence; for openers! Soon Google Milky Way will form and we will then do what is happening to us now; and that is; collect data of all types about everything for storage and access. The future is a library.

The Atom Pilots

I once imagined what it would look like if the Gods were controlling everything from a quantum particle base of operation. I imagined little jet fighters with cowboy like pilots bumping into and lassoing individual atoms into place to generate the events that only they understood. This, of course, would be the cartoon version of something far more likely to be a reality than boring gray aliens probing our rectums out of some demented cosmic perversion. That reality could be a deeper more exact control of events as orchestrated through ordinary matter, or … or maybe, maybe through our consciousness. If there is something special about the way consciousness happens then it could be a cosmic case of 'We don't know what we don't know'. But if there is quantum manipulation that is having an effect on our decision making, it may very well be something that we both need and have always had. We would effectively be quantum puppets. It then begs the question then. Does it happen in terms of a natural evolutionary process or someone or some things specific intent. We may be generated directly from alien intent. It means we may be, in essence, living an intent verb in action thinking the whole time that we are in control of our own destinies.

But there is no cause for alarm. On the other hand; what the fuck do I know? It could fucking be anything out there!

How slow the speed of light is

This is an important question that I don't feel is asked enough. In Hollywood astrophysics there is always someone there to potentially take over the galaxy like the Klingons or the Galactic Empire. It makes the Milky way seem so small. But in Hollywood astrophysics starships like the USS Enterprise from Star Trek and the Millennium Falcon seem to be traveling about the speed of light squared. For the record: It is presently impossible to travel faster than the speed of light with any known energy conversion process in this atomic dimension. But if such speeds were attainable then a trip to Proxima Centauri, our nearest stellar neighbor beyond the sun, which is 4.3 light years away, would only take about 10 minutes or so. This would be what would be necessary to create a galactic society and trade like we see in both Star Trek and Star Wars as well as many other sci fi presentations at mass communication birth on this planet. The present reality in terms of real world astrophysics states that a trip to Proxima Centauri, where you could stop and actually say hello, would take about 6 years or so. The additional years above the 4.3 light years to travel there would be due to acceleration and de acceleration of the craft containing human occupants. Obviously this would put a severe damper on any continuing flow television or cinematic story as by the time Captain Kirk reached any nearby planet he would be way to old to kick any serious Klingon ass and probably ready for retirement. Yes, I know that there would be a time dilation effect which would cause him to age at a slower rate than his human counterparts left on the Earth but in order to really get the real benefits of time dilation you have to accelerate at one gravitational level for a considerable period of time. Traveling to the nearest star wouldn't have that much of an effect. You really would have to travel thousands of light years across space mounting those decimal points ever closer to the speed of light to really reduce the aging process relative to the aging process on Earth. That being said, in the real astrophysics of deep space possibilities we are stuck with the incredibly slow speed that light travels. In fact, light travels so slow relative to our imaginative needs that it really should be embarrassed. To give you an idea of just how pathetically slow that it is: imagine one paper dinner plate representing the M31 galaxy or as it is known by many; The Andromeda Galaxy. Now imagine a smaller dessert plate representing our Milky Way galaxy. The two plates would be about 20 feet away from each other. If you could spend your whole life watching a ray light travel from one plate to another then you would never even be able to detect its movement over the course of your entire life because it would take about 2.2 million years to make the journey.

WHAT GOES UP NEVER COMES DOWN, SPINNING WHEELS...

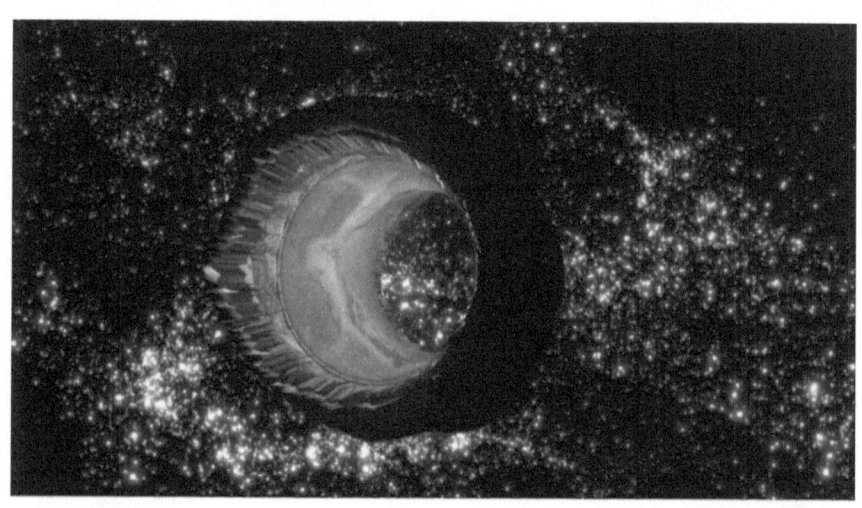

A centrifuge desert carved out from an asteroid.

Within the next cosmic second humans and dolphins will begin their journey inward towards the center of the Milky Way Galaxy. We will do so in starships of ever increasing size. To understand life aboard a starship you have to understand you would probably want to accelerate your ship at the same gravitational level as the planet you were born on. So if I were on a starship right now I would want to accelerate at one earth gravity and that way I would feel comfortable because I would be experiencing one earth gravity just like on the Earth. Once I had arrived at my destination then I could no longer accelerate at one G so I would probably want to assume a standard parking orbit inside the Goldie lox zone, which would be the precise distance from the star to allow liquid water to exist, and then spin a centrifuge and create artificial gravity that way. To understand this you have to remember those spinning cylinders you saw at the amusement park when you were a kid. Remember when you rode that spinning cylinder that pressed you against the wall and you thought maybe you might puke your guts out because of the three hot dogs you had just before. It basically the same type of effect with the hot dogs as optional; and a gigantic centrifuge spinning at just the right speed will produce the proper gravitational level for normal human inhabitance. Conceivably these centrifuges could be miles in diameter. These will be huge ships that will hopefully have mastered the art of complete self sustainment. All food, water and air must be completely recyclable. Every material, no matter how small, will be known and inventoried. Strict records of everything from soil to feces will be accounted for in precise numbers. If someone uses the restroom then toilet intelligence will keep an accurate record of the event and share the information directly with both waste management intelligence and even agricultural intelligence to best maintain an inventory of necessary recyclables. There will be no shitting around, literally. All food, of course, must be tracked; especially food not eaten by the inhabitants. Even this food must be placed back into this technologically constructed ecosystem for eventual recycling. And what will this environment be you might ask? Most likely on a deep space voyage between the stars the maintaining of the astronauts health will take priority. The distance between the stars is spectacular huge, making such voyages incredibly long. Most of these people will never come back. The idea of living in zero gravity for the rest of one's life would be ridiculous. Human muscles have a tendency to degrade rapidly in an environment where they are not exercised; and zero gravity would provide very little exercise; therefore gravity must be reinvented.

But who will inhabit them and where will they go? Starship astronauts will definitely be of a different breed than of anyone else. A mission to the stars will be a lifetime commitment. The astronauts will probably never return. To give you an idea how of far it is to another star, imagine traveling around the Earth seven times in one second. That's fast! Right? Now imagine traveling at that speed for 4.3 years to the nearest star system of Alpha Centauri; and that only represents the tiniest fraction of how far it is to the center of the Milky Way Galaxy of which is 26,000 light years away. In case you didn't know; a light year is the distance light travels in a year which is around 24 trillion miles (186,282 miles per second). Imagine traveling at that speed for 30,000 years. So obviously the distances are ridiculous; 'insane' may be a better word.

Yet within these insane distances lies a physics that may allow a single human being to make the journey into the center of the galaxy within his or her lifetime; and it is the Great Albert Einstein who predicted the now confirmed reality of the Theory of Relativity. Relativity states that objects slow down in time as they gain in mass as a result of increased in speed or velocity. Normally this would be considered a ridiculous statement made by a man clearly in need of psychological help--------except that its been proven to be true by so many different experiments that they could never all be counted. Relativity is a FACT! Mass does become heavier with velocity resulting in time dilation; and at speeds approaching the speed of light, mass increases considerably to the point where time slows down enough to allow individuals to make spectacularly long journeys and only age a few decades. According to Carl Sagan, if a starship accelerates at one gravitational level towards the center of the galaxy, an astronaut traveling 30,000 light years towards the center of the galaxy will age only about 21 years. Of course on Earth it is measured as 30,000 years. This is the effect that velocity has on ordinary mass. So if we sent an astronaut towards the center of the galaxy he would not get there until around the year 32,000, give or take a few thousand years; and in the process he would age only about 21 years or so but the Earth would age 30,000 years relative to his 21 years. This is called time dilation and it is what makes travel to other stars possible. The key aspect to it being; the longer you accelerate the more time slows down relative to what you are experiencing back on Earth. But who will go? And what will they be like?

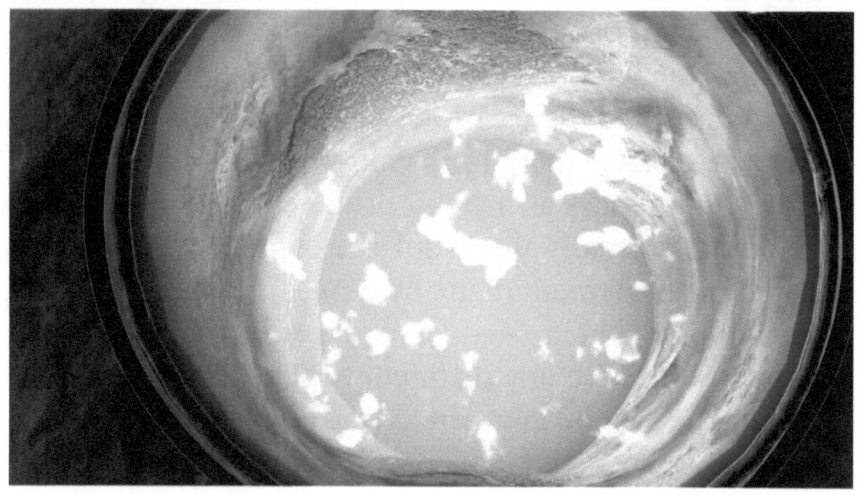

   As I said before, starship astronauts will be of a
different breed. A person preparing for a one way
journey to another solar system would probably have
to be well educated in a vast array of fields involving
astrophysics, nuclear propulsion, ecosystem
regeneration, medicine, and various types of
engineering. Every member of a starship crew has to
have extensive knowledge of everyone else's fields of
specialty. If something goes wrong, the idea of
radioing back to Earth for help, even if it is only for
instructions on how to fix a leaky toilet, is ridiculous.
That is why each member of the crew must be able to
back up any other crew member who might become
disabled. It wouldn't do any good to have just one
doctor or one nuclear propulsion engineer for the
obvious reason that if they become incapacitated then
that aspect of the mission would be in jeopardy. So in
order for a person to become educated in so many
fields he or she really must make a decision relatively

early in life that they would like to train for a position as an extra solar astronaut. It would be a process that would take perhaps fifteen years minimum to complete. So a person would probably have to begin training in their early adult life, because remember also, these missions are essentially forever; so the longer an astronaut can live, the better. It brings up all kinds of personal issues involving family dynamics that would be very difficult to deal with. Imagine telling your mother that you are training for an extra solar mission to another star system that will never return to the Earth. Imagine telling her that you've only got another 10 years or so to spend on the Earth whereas she will never see you again. That's not even taking into account the incredible dangers of the mission. It would be an uncomfortable conversation to say the least.

But all this training would be necessary because in space no one can hear you scream!

Earth may be somewhat limited in it's assistance to a troubled starship thousands of light years distant

In space no one can hear you scream for customer service that is;

Customer service in deep space isn't it all like what you might find in one of your local retail outlets. I mean here on Earth you can get an oil change plus a 14 point inspection for only $19.95 with coupon through the month of June. But in deep space if you suffer an asteroid strike or catch fire you could literally rewrite the book on what it means to be screwed.

Let us now remember what it is that we are talking about here; let us now remember the vastness of deep space. If I were to send a signal to a starship one hundred light years away, it's gonna take one hundred years for that signal to get there. If they wanted to transmit a "hello" back to us it would take another one hundred years for a return reply. I mean my great grandchildren are gonna be waiting for that "hello". It's difficult enough to get kids to take a message anyway. Think of it this way, if a signal was launched when you were born and now; however old you are; 30, 40, 50 or whatever, you finally get the return message then you can get a better feel for the vastness of these immense periods of time. The trick here, of course, is to use the one huge time period that you know best; your age. That's why I always said that when you launch people to another star you can pretty much kiss their asses goodbye forever. You're never going to see those bastards again. Every once in a while you might hear a scream but that's pretty much about it. Or you might get a message back letting you know that food is low and we're eating each other now; or worse yet, if you were to send an SOS back to your home planet, it's going to be received by a civilization that may have forgotten that they ever even launched you or worse yet may not even give a shit. Can you imagine sending a message out when you are a grad student and then waiting your whole career and then finally getting a reply back when you are eighty saying *"The number you dialed has been disconnected please dial again"*. More realistically, can you imagine getting a message from a starship seventy five light years out that they have suffered a centrifuge fire? This would mean that by the time you got that message it would be seventy five years later. And it might say something like "This is Captain Gullible, former Penn State quarterback; now Captain of the Starship Titanic. We've struck an asteroid. I'm sorry that I ever let you people talk me into this shit". And the best that you could tell him was that you are sorry to hear about the fire but tech-support switched to Nigeria 200 years ago. Here's their number. In short, in deep space you are about as much on your own as you can get. But the good news is that you won't be alone.

Field mice and the center of the galaxy

One of the most fascinating evolutionary forecasts of the future may be that of the ordinary field mouse. Field mice are notorious for getting into everything; setting up residency in virtually anything from houses to trucks to ships to planes. There's no doubt it, the little bastards get around; and no doubt at our birth of technological construction in outer space field mice will be more than happy to assist us in our endeavor .

Yes, beyond the great orbit of mars, beyond the mighty moons of Jupiter, the little bastards will be there. The field mouse will follow us into the beyond; light years into deep space toward the center of the Milky Way Galaxy; and what it will evolve into, I have no idea. But in consideration of the different simulated gravitational levels that will be produced in centrifuges in outer space it is likely that the evolution of the field mouse will involve an accelerated evolution of hand/foot motor dexterity. This is because in outer space we will construct huge rotating wheels to simulate gravitation. And how fast we rotate them determines the gravitational force. Some people will perhaps prefer to live in a 3/4 gravitational spin. Others will prefer a 1/4 spin allowing them to fly around and play Spiderman whenever they want. Field mice, being as light as they are, will plummet everywhere in these environments; yet a certain stability is required for survival. Therefore only those field mice equipped with strong hands and feet for grabbing and maintaining positioning should survive. It is important to understand that this evolution of hand motor dexterity will involve a significant advancement of intelligent perception. Early human ancestors such as Proconsul, which lived some 40 million years ago, experienced rapid cortex development which may have been stimulated from the need for complex hand dexterity for life in the trees. And this may also happen for the field mouse of tomorrow who stows away on an interstellar flight. Field mice will come to know a greater degree of hand potential for grabbing and climbing and, as a result, the evolution of thought will increase as well. What they will ultimately become I do not know. Yet they are very resilient and extremely adaptable and most likely will survive. Yet it is an interesting speculation. Will they really evolve into the center of the galaxy? Might they actually someday evolve a conscious awareness inside some huge multi formational computer hard drive deep in the depths of outer space? Will they perhaps wander where they are and how they got there and perhaps think that the Universe occurs in a huge technological formation of some kind in all directions? Or will they know that they have evolved deep inside the environment of some super conscious technological god (us) and then begin to make their way upward toward the surface, in an infinite array of technological landscapes in all directions to say "hello", we are conscious of you.

As mankind moves out into the stars his starships will no doubt become bigger and better. Advanced digital intelligence combined with equally advanced robotics will allow faster and faster construction of larger and larger ships. Soon human beings will be dealing with distances well beyond a mere one or two light years and instead may be dealing with distances of 10's or even 100's of light years. And no matter how advanced these ships become they probably will never be able to exceed the speed of light. This means that a ship that reaches its destiny solar system that is hundred100 light years from Earth must still wait hundred years for new information to arrive from the Earth. Think of how many technological advancements will take place in just the next hundred years. It would be like if you were stationed hundred light years away from the Earth with the technology of an abacus as your most advanced piece of hardware; and on the Earth they just developed the Cray Supercomputer; and now the people of the Earth are going to transmit the information to you on how to build one. It is still going to take another hundred years for the instructions to reach you because nothing travels faster than the speed of light. By the time you get instructions on how to build a Cray Supercomputer, it will long since have become obsolete. Eventually, there will be situations where a starship eventually reaches its destiny near the center of the galaxy some 20 or 30 thousand light years into the future. This means that because they are on a starship their technology will only advance so fast because of more limited resources and time allocation. But by the time they reach their destination 30,000 years into the future the people back on Earth will be 30,000 years ahead in technology; 30,000 years of technological advancement is inconceivable. But if those people of Earth were then to leave for the center of the galaxy it will still take them 30,000 years to rendezvous with this older outdated ship.

There no doubt will be situations where one starship will rendezvous with another starship that left the Earth a considerable amount of time before and find the occupants either dead or dying; of which it will then spring untold technological advancements on the elder starship beyond their comprehension. It would be like if Columbus reached the new world in his 3 wooden vessels and then another ship called the U.S.S Ronald Reagan Aircraft Carrier arrived from Spain a few years later. Or if somehow the speed of light were circumvented then the U.S.S Ronald Reagan Aircraft Carrier would be waiting for Columbus's three ships to arrive at the center of the galaxy. This could actually happen if new forms of exotic matter found near the speed of light were harnessed to actually surpass the speed of light. Regardless, the technological advancements would be vast from one mission to the next.

Ultimately starships will probably have to unite in deep space to survive. Deep space is a cold, vast, unsympathetic eternal abyss. The problems that can take place cannot be conceived. A lot of things can happen on a mission to the stars. The ship could start to lose air of which even 1/1000 of a percent a year would become a serious problem. Radiation or cosmic ray bombardment could affect so many different things that I can't even begin to list them. Plus the fact that there are so many unknown electromagnetic phenomenon that exist outside of our solar system that are not understood in any significant way. But through all this, starship design and especially the timing of launch, will be crucial in maintaining support and survival for all involved. There may, however, come a time when, say, a starship no longer receives support from the Earth, say, because of a war or maybe an economic collapse on the Earth. Whether or not a starship could actually survive in space and reproduce indefinitely into the future is, I guess relative to how advanced it is. Ultimately, one has to assume that a ship will begin to lose its resources slowly over time; it is unavoidable. However the starship that can harness various forms of matter and then manipulate them into what it needs may survive. This no doubt will become the hallmark of survival for those who choose to travel to the stars and beyond.

A rogue interstellar alien examines a long forgotten centrifuge zoo of Triassic Earth in deep space abandoned by his ancestors thousands of years earlier.

Ultimately in the end we will probably find massive orbital graveyards of spinning centrifuges in space. They probably litter the intergalactic core as we speak. These ships would be burned out, frozen over, abandoned broken down ancient monuments, slowly rotating around the center of the galaxy over millions upon millions of years. Most would be traveling at spectacular speeds but not accelerating. There could theoretically be some occupants left; possibly even second or third generation people that would be like Eskimos having no idea how they got there or what their whole round ringed world is. They would have no knowledge of the Earth and would essentially be traveling through deep space uncontrollably until they collided with a star or something else.

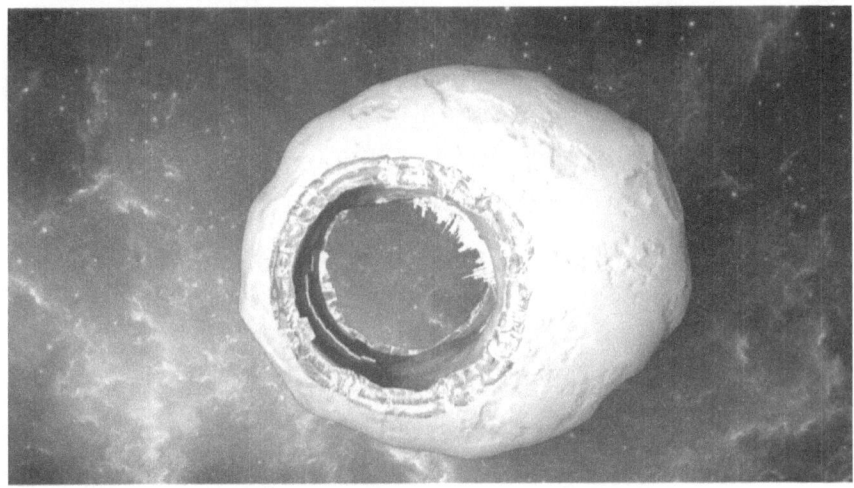

A frozen over space centrifuge that was carved out of an asteroid lies abandoned in deep space.

Does the increase of energy nearer to the center of the galaxy aide higher level technological species in galactic travel.

I have always had the sneaking suspicion that this atomic dimension is somehow constructed to ensure the survival of the higher forms of intelligence if those intelligences can survive the onslaught of constant death and destruction that actually generates the evolution of these higher intelligences. That being said: It may very well be that all of the atomic synthesis of matter that is necessary to pave the way for a starship to make it's way towards the center of the galaxy to ultimately enter into the Sagittarius A black hole and become whatever the wonders of atomic dimensional transformation may behold may exist. There is a tremendous increase in plasma and fusion output as we move closer to the center of the Milky Way Galaxy that may be necessary to allow a starship to make it's way inward. Corona mass ejections, which are massive electromagnetic outpourings of super heated plasma, could theoretically be harnessed by deep space starships for continuous fuel consumption assuming that accurate long range analysis would allow for accurate predictions of Corona mass ejections. We may find super arc interstellar trails made of Corona mass ejections that a starship could first predict and then time a gravitational passing by to then reach out somehow and capture the outpouring of a Corona mass ejection plasma for life sustaining energy for not just propulsion but all other energy necessities as well.

Digital record erosion

But some starships may be successful in mastering
regeneration and surviving; and yet through all this,
our image of the Earth may still disappear. Our deep
descendants may eventually lose their knowledge of
the Earth as they travel ever deeper into the distant
future or as stated before, the infinite past. Infinity is a
long time; and their records, in accordance with the
laws of physical breakdown of the Universe, must
disintegrate into the distant future as time goes on;
unless, of course, they can actually see into the past;
then, of course, maybe throughout the depths of
infinity they may remain aware of the Earth, which is,
of course, their birthplace; otherwise there is no
physical way to stop the degradation. It is a
fundamental law of the Universe. All our records,
movies and tapes and cd's and whatever else we may
have encoded on our most digitally advanced
superstructures of the future will not survive. Even
with the most advanced laser and computer chip
documentation of the distant future, the slightest
alteration of any kind, which always must happen,
will alter the documented structure. Even if it is only
one billionth of a billionth of a billionth of a percent,
infinity will beat you. Infinity will slowly watch it
disintegrate into dust. Even reconstructed copies must
alter. It's depressing to think that someday the original
Star Wars will someday be lost from its original form.
What does all this mean in terms of Evolution you
might ask? It means that traditions of life and art must
eventually die but somehow the overall parental flow
that contains it survives. Humanity will survive but

it's traditions and arts will eventually die or be altered unrecognizably. Evolution is a brutal_thing; it is beyond most people to accept it because of it's unforgiving nature.

Life in another dimension

Only about 3 cosmic hours ago did we discover technology in the jungles on the Planet Earth and now in about 40 cosmic seconds we will be approaching the center of the galaxy upon which we shall discover the next great super direction in our physical motion through atomic substance acceleration into another dimension. It should obviously be understood that we are not defying the theory of relativity. We are simply following its master teachings in a natural universal evolution. And let it be understood that the theory of relatively does not state that it is impossible to travel faster than the speed of light; it states it is impossible to travel faster the speed of light *with any known energy conversion process in this atomic dimension.* Yet matter seems to morph into energy on its way to the boundaries of this atomic dimension at the speed of light. Yet under the ungodly pressures exerted in the centers of massive black holes, which are by far the greatest pressures exerted in the Universe, light and energy themselves are transformed to just outside of this dimension. And while I do honestly believe this does occur. I must wander; is it possible to be alive in an atomic substance different from our own? Are there laws of physics that are compatible to what created us here that can sustain us there?

The Universe has, at this point, ensured our very

evolution in a hauntingly implied direction to other possibilities. Stars are allowed to form; planets can revolve around them in a stable orbit; amino acid formations form and are then capable of reproducing a natural selection growth of evolution betterment. Technological physics is directed to place these beings into outer space above their planets and the phenomenon of time dilation of matter is, of course, waiting to allow passage to the stars. And, most likely, in the direction of the time dilation of matter, mass metamorphosis of living beings into the next dimension is directed in some way to maintain the existence of life forms formally created by this atomic substance of which we currently exist; but it only allows the survival of the fittest to go. Whether or not it is possible to make this journey beyond the speed of light naturally or whether special technological preparations are needed to get there I do not know. And ultimately in a super distant evolution, the eventual finding of another distant dimension existing in a far off dimension beyond our own will introduce us to physics beyond our present comprehension. What will be found there? What are the laws of physics beyond the speed of light? What is the difference in the evolution of events in experiencing this kind of phenomenon? Would we be beyond our conscious knowledge of our natural understanding of nature as we know it? These are just some of the questions to be asked. The laws of physics in another dimension would no doubt be different, however, there may be some basic similarities all dimensions have. A basic example would be the domino effect. It about as simple a concept as one can imagine; and it probably exists elsewhere in the laws of physics of other dimensions. That might be as much knowledge as we get before we enter. It is probable that in a journey outside of this dimension one may have no senses at all. The physics of occurrence would be beyond our understanding. Vibration and sound would at least be spectacularly different and who knows really what, if any, senses would work in another atomic dimension. Most likely technological civilizations at the center of the galaxy would have a far greater understanding of the physics and mass metamorphosis phenomenon that may take place in other dimensions. They know more of what will happen and most likely, in a huge evolution equation, will realize the knowledge of mass metamorphosis evolution from the knowledge it took to get them to the center of the galaxy in the first place. It may very well be that their biological senses must receive all knowledge and perceptions through huge artificial intelligence constructions sustaining them with knowledge about this new dimension and it also may even be the that biological beings themselves may have to be fed into an artificial intelligence for their own sustainment. It is difficult to conceive what may be perceivable immediately outside of our atomic dimension. Most likely the centers of massive black holes could be perceived on the outside of our dimension. Perhaps, as Carl Sagan has stated before, they are white holes. Or maybe they are shining stars in the heavier dimension ahead of us. Wherever black holes lead, it is important to understand that we may not actually be in another dimension immediately outside of our own. If we were to travel around a black hole we may find ourselves at the edge of a long funnelization between two dimensions; it could be the beginning of a long funnelization into another dimension, not through spatial distance, but through universal atomic substance change; and it may very well be that like the great distances between the stars, the distances between two atomic dimensions could be vast beyond belief; so life must exist between these two dimensions, as well, if it is to survive the velocity increase direction into another dimension. As a matter of fact, it could be that as we transform our matter into another dimension that we never really receive knowledge that those who have already made the trip survived. It may be that it is just predicted that we would survive. So theoretically, the entire human race could go extinct by traveling into another atomic dimension because we would never know for certain that anyone is surviving the trip, or transformation. So the question remains; could the physics of another atomic mass dimension support life? If it could then most likely our far off descendants will find themselves the equivalent of bacteria in this new dimension; totally blind, deaf and incapable of anything. Even the simplest physics understanding in this new atomic dimension may be millions of years beyond their most primitive comprehensions. These laws of physics will most likely be billions of times more complex than our own primitive dimension. Will we suddenly become the equivalent of ants in relation to beings that can build high speed computer processers.

And in terms of a cosmic perspective by the time we reach a comfortable comprehension inside this far off dimension, the chimpanzee may have evolved a sophisticated technological comprehension in orbit around the center of the of milky way galaxy. By then the Earth will be another place. Many of the animals we now observe and watch on the cable television will now be evolving into a spectacular awareness of the stars and the Universe around them. Maybe the white shark will become an intelligent lover of environmental beauty and become a global protector of the natural inhabitance of the Earth for future species evolutions.

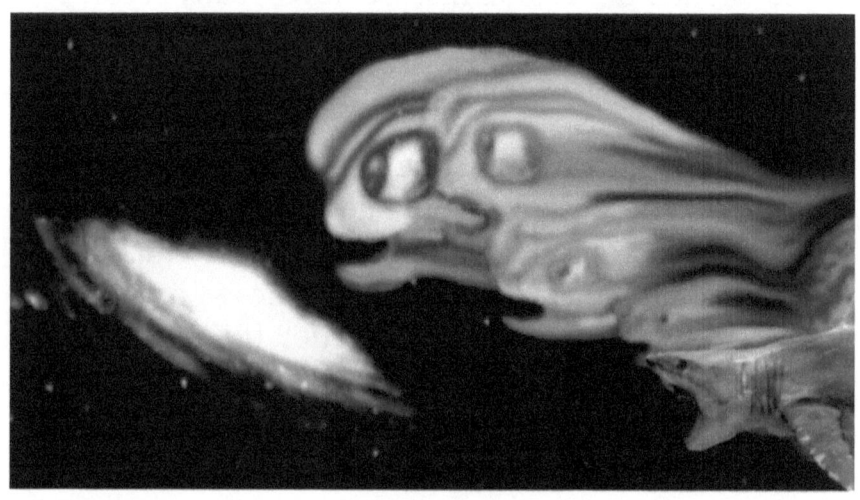

50 million years from now the Great White Shark may evolve into something beyond our present comprehension; or perhaps, with the help of extraterrestrial intervention, it already has.

Or perhaps a technological intelligence from another eon will come and inhabit the Planet Earth on their journey through the Universe.

Something like this may eventually happen; or

perhaps something like this has already happened; some sort of extraterrestrial visit is a probable event in the life of the Earth. And while I will not speculate on what it may have been, the cosmology involved may have produced the birth of new directions in the evolution of life to and from the Planet Earth; perhaps well beyond the Planet Earth as well. Observational experimentation for super conscious analysis may be the reason to abduct species from this planet. No, I don't believe little gray men from wherever are abducting people and having sex with them. I believe instead that the people who are claiming this may be having sex with themselves unwittingly.

It may very well be that at our birth of mass communication technology we have signaled the Cosmos of a new birth of technological conscious awareness in space; a tiny little planet is now radiating it thoughts and emotions in all directions from its cradle, poised deep in the depths of space. From Howdy Doody to late night with David Letterman to CNN evening news, we are screaming like a new born child into the cosmos. It is the insanity of the hydrogen atom left unchecked by the Gods for billions of years that has allowed consciousness to prevail; and now we are blasting our consciousness outward in all directions; and only now at this exact second in deep space have some of the great intelligences of the Universe detected an awesome explosion of a global technological mass communication exploding in a single second from our planet. We are an explosion in the Universe screaming our birth of conscious awareness into the cosmos at the speed of light. Cosmic intelligences from all directions will turn in our direction to look and many will investigate.

Mass Communication Birth is a sudden explosion of self awareness into the Universe and even though signal degradation is significant in the first few light years out producing significant loss of data the existence of these transmissions may still be detected much further out.

THE NON-TECHNOLOGICAL GODS

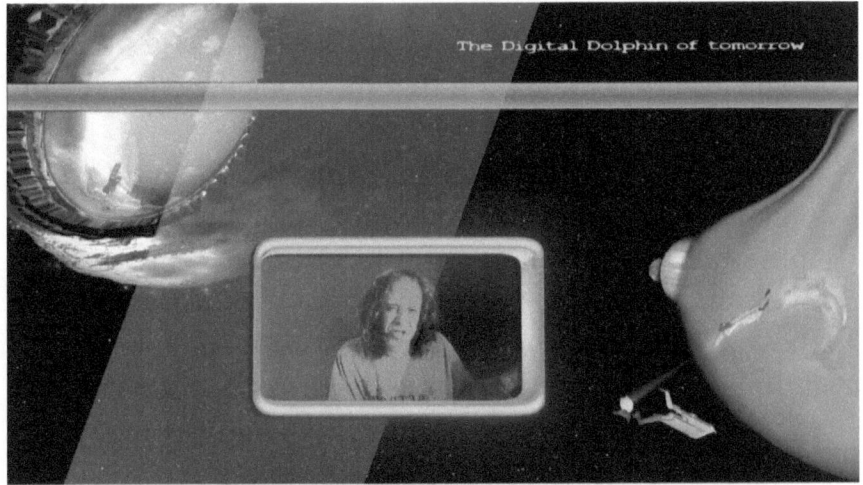

The Digital Dolphin of tomorrow

The hydrogen plasma that became this atomic dimension billions of years ago has now assembled itself into consciousness. And as man and woman step out of their little cradle on Earth and into the great big black Universe of darkness they will discover that there are beings hundreds of thousands of times more intelligent, that do nothing but talk to one another; beings of superb intellect, far beyond humanity in conscious evolution, who are incapable of grasping the pencil, yet still explore the passions of conscious awareness beyond homo sapiens comprehension. And it is here that we have the key from which the simplest and most powerful passion of the Universe stems; socialization from sexuality; the great fascination of all intelligent beings in the Universe; one another. Socialization from sexuality is, of course, an aspect of Evolution that will always take place to ensure the survivability of offspring. It is the great super exchange of one another all stemming from the need of sexuality which in turn stems from the universal occurrence of natural selection. It is the evolution of emotional exchange that produces higher and higher levels of communication that provides the edge for the survival of the fittest.

Yes, the great beings talk to one another in the Universe and they love to explore one another's minds and bodies in an infinite and never ending evolution of sexual and comic odysseys. And while they are unable to stand a domino on its end, their communicative consciousness will eclipse the human race by hundreds of thousands of years. It is we who will follow them into the Universe and it is we will come to understand our technology as their toys of play by which their conscious understanding of the Universe will flourish. We will construct their technological senses for them to view the Universe in a way completely alien to us. From far out telescopes to inner spatial microscopes; these technologies will be outside of our use as human beings. They will be built for beings vastly alien to us. But yet it is we who will build them. It is we, as human beings, who will assume the role of technology masters as well as technology slaves but, of course, everyone will be working for everyone. And over the course of large scale galactic time it may very well be that there are beings in the galaxy who must make their way towards the center of the galaxy and it is our job in evolution to get them there. Why, you might ask then, would these beings, perhaps, be more intelligent? Time may play a factor but more importantly it may be the relationship between their medium of communication and their medium of perception. Human beings communicate with sound yet perceive with, primarily, their sight. But there may be alien intelligences that communicate with the same medium of both communication and perception. This would give them a tremendous advantage in receiving and comprehending information. The relationship between what is being perceived and then communicated is a far more exact match because both mediums of perception and communication are the same. Human beings have to transform from one medium into another. In other words, we must transform light perception into sound communication. This, in essence, may be a handicap in an evolution Universe. But always remember we are the technological constructors of the galaxy. No matter how intelligent an alien being becomes it may very well need us to, not only travel from one place to another, but also to build its technologically constructed senses to explore the Universe with; and in exchange we will receive an abundance of information in forms unimaginable at this time. We may very well be the semi-trucks of the galaxy and they, meaning the non-technological intelligent beings, may very well be the brains of the galaxy. And even though they are not about to develop technological progression, they still have chosen other equally and perhaps more powerful evolutionary directions on which to embark. And in the course of this cosmic study we shall examine one such awesome creature; a creature of superb socialization evolution; a creature with a huge massive convoluted brain and a communication systems network still beyond mankind's comprehension at homo sapiens Mass Communication Birth.

Driving Miss Dolphin

In terms of peoples perceptions of aliens; people like aliens that are different yet similar to us in a way that we can relate. In other words, we wouldn't want anything we couldn't potentially fuck. Generally, I divide aliens into two general classes. Right now you have your technological aliens; that would be us; then you have your non-technological alien intelligences such as the bottlenose dolphin of which isn't an alien, of course, but whose intelligence is so radically different form our own that I would be willing to bet that it would be more different than any grey alien; if they existed. They would be an excellent example of a non-technological intelligence.

For as long as recorded history can document, dolphins have consciously aided in the survival of the human being. In their huge brains exists the compassion and understanding and comprehension of need beyond animal utilitarian tasks. The conscious ability to save a drowning child or come between a shark and a man and provide conscious escort to assured safety requires thought far beyond simple utilitarian consciousness. It requires a consciousness able to see a given situation in a variety of options as well as a being who is able to conceive in alternatives beyond basic animal instinct. This means that it is a being of which is not only required to think and analyze but also to feel the compassion to leave its natural norm and act upon it. These are thoughts beyond the utilitarian instinctive mode, beyond the mere animalistic need of food, offspring, and self survival. These are the thoughts of a being in conscious analysis of its environment.

Dolphin brain reality

The brain of a bottlenose dolphin is some 200 cubic centimeters more massive than that of the human brain and nearly twice as convoluted. It is a brain of intricate complexity containing some 2 and 1/4 times as many neurons as the human brain. And while this is a brain whose complexity is poorly understood, we must first begin by understanding that it is a brain of completely different environmental evolution. It is a brain of what I call "duel medium consciousness". Which means it is a brain that perceives and communicates with the same medium; sound! Because of this "duel medium consciousness" and because the neo cortex is so developed, it is reasonable to assume that they have dealt in concepts and ideas that we have never even conceived of. This is, perhaps, a prime characteristic of an aquatic being. And while this is an important concept in terms of alien conscious perception, the real importance is the actual cognitive and perceptional senses themselves.

Now occasionally you will hear someone say something like 'human beings are the only ones who could do this or we are the only ones that can do that'. People like to bat this shit around like a nerf ball. We think we're special because we have opposable thumbs. But having an opposable thumb doesn't do you any good if it is shoved up your ass. So I've got news for you; we may be number two when it comes to processing information. Remember, just because we are these opposable thumb manipulating bastards doesn't necessarily mean that we are the Universes secret favorite. The fact is that the dolphin does appear to have the more advanced brain in terms of cortex development. In addition to that they do have a greater neural density and a neural connectivity plus the large silent areas of the cortex for reasoning and complex thought are extremely developed.

To understand the true measure dolphin intelligence one must understand the relationship between perception and communication. As I said before, in human beings most perception is accumulated though light radiation intake. Human beings are generally visual animals who gain about 80 percent of their perception through their eyesight. They must then translate that medium of light radiation perception into an audio communication. The transference of perception to communication using two different mediums will always be an impediment in an intelligent beings communicative evolution. But the dolphin mind uses the same medium (sound) to transfer perception into communication giving it arguably a huge advantage over homo sapiens in the communication of information; and it may very well be that this particular trait defines the hierarchy of truly intelligent beings in the Universe. It is the non-technological intelligence that is incapable of grasping a tool; yet because of its perceptual communicative evolution, meaning its ability to perceive and communicate with the same medium, it may very well command the tool users of the Universe with its own intelligence, even though it is the tool user who maintains the militant upper hand. Why?; because the non-technological being that perceives and communicates with the same medium has a higher communication threshold, meaning it communicates a more accurate record because it does not have to transpose mediums. Think about it; if I describe the way someone looks to you; unless I compare it to someone else who looks similar or start listing visual traits of that person; you are only going to get a very general picture of what that person looks like. But a being who sees with sound then communicates with sound may very well transmit a higher more detailed version of what they are describing because less is lost in the translation. A dolphin may very well be able to communicate an exact sonic visual memory of something to another dolphin. Human beings have to take a picture to do this. And it is here that we may have a possible understanding of the homo sapiens true role in the Universe; that of matter manipulator; a tool creator capable of manipulating the Universe to the benefit of himself, but more importantly, other intelligences. And while human beings may have the ability to create and destroy the vast majority of life forms in the Universe, it is far more likely that we will provide the manipulation of matter into tool in exchange for the receiving of information from higher intelligent beings in the Universe that are incapable of tool use.

In terms of the dolphin brain, it is an aquatic medium consciousness and the fact that it uses three-dimensional sonar allows it to perceive and relay its sonic perceptions of the shape and elemental matter makeup to interpret the internal structure of its environment. Because of this evolution of perceptual communicative similarity, the dolphin is able to know and communicate things that a human being could never be aware of. In other words, it may be a more mathematically exact conveyance of information. Think of it like this; imagine that you have five speakers under water placed at different points in space and time. As you can see below, they are in the shape of a pyramid.

And when they all sound off then you would actually hear the shape of a pyramid as opposed to actually seeing it.

The mysteries that dolphins must contemplate

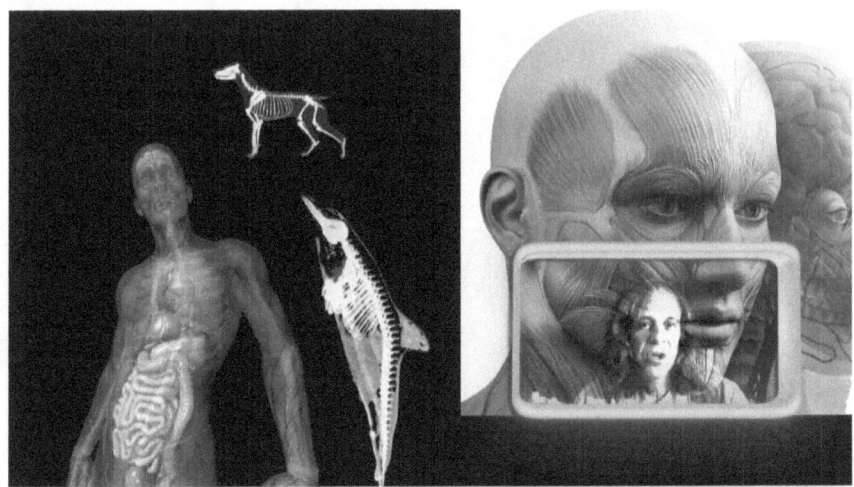

I love imagining in my brain what might be going on inside their brains. For instance; if a dog jumped in the water with a dolphin then the dolphin can sonar its body to get a three dimensional view of the dogs skeletal makeup. The sonar beam emitted from its head will reflect off the bone structure and ribcage of the dog and return a sonic image back to the dolphin; the dolphin can then see with sound the internal skeletal structure of the dog and because of this it knows that compared to its own skeletal structure, they are very similar. Sonic analysis of a sharks internal structure, on the other hand, would reveal very little similarity. This is because the sharks physiology is largely made of cartilage which would be of a different design than a mammal; so it knows it is more related to this little hairy thing that lives on land than it is the shark of which it looks generally similar to and also lives in the water.

As we can see above the view that a dolphin has would reveal an internal structure of other anatomy that it could compare with it's own internal structure.

Likewise if a human being enters the water, the dolphin can see the human beings skeletal structure and clearly see the similarity of the skeletal structure that dolphins and humans share. It knows they both share a similar biological construction indicating perhaps a similar common evolutionary past; thus the dolphin then knows it is more related to the human who walks on the land then it is to the shark which swims in the ocean. They may also have related human technological display with human hand dexterity as interpreted by their own sonic analysis; and because of this they may very well be amazed at our ability to create vast arrays of technology out of nothing but rocks and mud in the Earth; and they would not be incorrect if they did so! They should be able to do this because they know that what we create technologically is related directly to the matter they encounter in their environment because they do, in fact, recognize similar acoustical chemical properties found in the Earth with those of our technological constructions as determined through their own sonic analysis. Experiments have shown Dolphins to be aware, through sonic image, of chemical differences in a wide variety of substances. And if so they must be utterly baffled about how homo sapiens could mold or manipulate an environment of a lot harder substance than their hands are made of. And because of their improbable and obvious lack of knowledge of fire it must truly be we who are perceived as alien. What must they think upon their first view of a perfect right angle fissure or the curvature of a perfect geometric sphere from solid concrete or steel? To them we are a study in alien wonderment; an evolution of cosmic fascination. To them, we are a completely different unknown direction of consciousness. This must be the topic of great discussion among dolphins as to the common evolutionary heritage of which we, meaning both man and dolphin, both sprang.

Dolphin addiction

As we know from being human; human beings can literally become addicted to anything. There is no level of which we will not sink in our obsession for the next ridiculous flavor of the week; and dolphin interaction shouldn't be any different; and it may even be that we may actually become dependent on them someday in terms of the processing of information as well. I know what you're thinking; "that's ridiculous; us depending on dolphins"? But remember when we become dependent on things, we allow ourselves to become dependent on things for convenience. It's like depending on a blue tooth for your phone. We don't really need it to survive and I grew up without one, of course, but I'm not really sure how I made it that first 48 years without one. Just the other day I lost my Bluetooth and I panicked. I had become dependent on something that I had survived without for decades. Not having it should have killed me, but because I didn't know that; I survived for 50 years without one. But being dependent on technology is, perhaps a uniquely human trait. Nobody ever accused me of being a control freak; and I'm actually a pretty laid-back guy, but by virtue of the fact that I'm a human being; I'm a manipulating son of a bitch. I'm looking around this room that I'm in here and everything that I see is manmade or sculpted out exactly the way I want it to be. Having grown up in an environment where they've never had any control, the dolphin will see things differently. There may be a greater ease of acceptance of events because of the total lack of control over the environment. To give you an idea of the difference; if a hurricane is coming, human beings may choose to evacuate or they may choose to stay and have a party inside the hurricane of which has happened many times before with disastrous results. It's that stubborn sense we have about control or in this case the lack of control and our defiance to it. We like to control everything because we are the out of control freaks of the Universe. But a dolphin is smart; it will get the hell out of there.

Yes, sometimes people like to press the buttons of nature just to see if the Universe is really serious about killer hurricanes as we see here with the hurricane party happening in the middle of the hurricane.

Eventually, as our cultures and societies integrate, the control freak and the non technological god will do their very best to co exist but understanding the intricacies of each side from the outside looking in will take longer. This is because the structural dynamic of human and dolphin mindsets as determined by the environments of which we both have evolved are radically different.

Life is a ballet

Dolphins live in a motion communication society of emotionally charged conscious synchronicity. This means they are aware in consciousness and communication of one another's motion and position in space and time. To understand dolphin consciousness one must understand the ability to hear sonic images in a 360 sphere from all around. And even though a single dolphin can only scan 20 degrees in a single burst of sonar, he can probably easily pickup the sonic images of other dolphins as well. By this I mean he can easily pick up the reflected sonic images of all the other dolphins that are behind him, below him and in front of him. It is as though we could see the entire 360 sphere around us in all directions. In comprehending dolphin consciousness one must understand that it is not only communication with sound, it is communication about sound that dolphins engage in. While human beings receive the majority of their information from sight then communicate that information in a sonic form; dolphins communicate in the same medium of which they perceive. Therefore it is reasonable to assume that whatever syntax dolphins may have evolved may be frequency relative to the echoes that they receive from certain objects. They may very well engage in quite sophisticated discussions of sonic analysis. These conversations may involve complex exchanges of frequency cycles of the absorption of echo bounces and sound classifications that may involve a conscious mathematical analysis integrated into language. As a matter of fact the relationship between dolphins and mathematics may indeed be intricately integrated into one another not only in terms of sonic analysis but also in the form of motion symmetry consciousness. This type of consciousness refers to the ability to not only be aware of one another's position in space and time but also to interpret the sonar analysis of an object to produce a particular sonic conclusion. In other words by combining sonar beams of different frequencies off one object in synchronicity, a unique sonic knowledge may be gained. Baby dolphins, immediately upon birth must begin a motion awareness of their mothers, therefore integrating this type of awareness and/or communication at an early age. This is necessary to lead to a greater communication with other adult dolphins. Baby dolphins may even perceive the external world from the womb through sonar? There may even be a sonic communication that takes place between mother and child before external birth?

With dolphin sonar being emitted everywhere in a school of dolphins it may be possible to know the location of virtually every member of the school. In an actual possible interpretation, dolphins may think, not only in their own sonically returned images, but also in the actual sonic images of other dolphins; thus producing a possible three dimensional spherical view of their environment and of the other dolphins. This very well may happen because in the water their sonic emissions echo and bounce off virtually everything in all directions; thus the returning images can most likely be picked up by other dolphins which then allows a perception of a view normally unattainable by a single dolphins own sonic emissions. This is an extremely difficult thing to imagine and opens up some fantastic possibilities because not only can dolphins receive images of the internal and external structures of objects in the three dimensions of the spherical space around them, but they also have evolved a wonderful sense of play within it. They may, in fact, have even come to enjoy some immensely evolved pleasures from the viewing of one another's own sonic perceptions? They may have even evolved into the art of viewing and feeling one another's sonic motions in space and time and feel and exchange sonic pleasures involving an infinite array of motion configurations that they have formed around one another? And for every dolphin to see the sonic images of every other dolphin would create a societal dynamic that humans beings are now only beginning to create for themselves through video transmission at their birth of mass communication. In dolphins this may exist because dolphin evolution in the open ocean has required the maintaining of a constant food supply, which means the eventual communicational need for precise positioning and timing in the gathering of the fish has become a need that has spawned the eventual birth of this particular consciousness in the galaxy. And like all evolutions in the wild, necessity is the mother of invention. And the great communicational pleasures intelligent beings enjoy in the Universe have openly evolved from a primitive need for survival in the wild. It is the continued evolution of a communication media to match the continuing evolution of a prey; and as dolphin strategic communication improved, only the quickest of fish could escape thus surviving to produce only the quickest offspring requiring the dolphin to evolve its communication and eventual awareness still even more in evolution. It is the never-ending "chase" of evolution; because when prey evolves then predators must evolve forcing prey to evolve again and it will end only when the Universe ends.

To understand or feel the actual depth perception of dolphin motion symmetry consciousness is, of course, beyond human comprehension. And while this may indeed seem a bit far out, most likely their evolution of sonic communication did evolve from the need for communicational strategic positioning to catch fish in a three-dimensional environment. If this is true, then it is likely a majority of their communicational conscious awareness does revolve around the sonic knowledge of the relative motions and positions of each other that they exchange during their play and hunting activities. They are aware of one another's position in space and time. However in the evolution of truly intelligent beings, which I believe the bottlenose dolphin and perhaps a few others have become, an actual conscious exchange of psychology has taken place. And even though this exchange has ultimately evolved from, say, a sonic motion and positioning type awareness, it is also tied to a communicational evolution in a three-dimensional environment. In the dolphin evolution of this communicational consciousness, an exquisite communication of timing, perhaps in complex mathematical notation has evolved. What intricate communications contained in their fantastic brain complexity could they be saying? It is easier in speculations like these to first understand the vital communications necessary to carry on day to day activities of survival. And although I believe dolphins have long since evolved beyond simple communication of survival and into a wonderful evolution of play, it is much easier to begin with what would be a more obvious need for communication.

The Dolphin Erotic

In attempting to comprehend dolphin consciousness it is crucial in our primitive understanding to acknowledge that in our own emotional growth we thrive upon the physical affection of each other. Scientists have shown that in humans and apes physical affection from the mother at birth is crucial to the overall psychological development of the young. Human beings cannot developed properly without it. The dolphins seemingly incredibly emotionally developed consciousness also appears to have evolved a central need for an actual physical affection since they spend an incredible amount of time caressing one another. But sonic sonar impact upon one another's bodies also may provide sexual communication in another direction of consciousness. It may be that dolphins experience fantastic sexual pleasures from the generation of sonic blasts of energy at one another's bodies. They may possibly even be focusing in on the more sensitive areas for erotic pleasure with sound and sonar as well.

Sexual sonar, if you will, sexually focused at variable frequencies into the body may produce vibrant ecstasies. This, of course, would be fantastically different from anything human beings could understand. Human beings sexual conceptions are spread for the most part throughout the five senses; none of these senses include a sonic impact upon the body. Instead human beings have, in large part, a technologically constructed sexuality. Our cloths that we wear and love to take off at various paces for sexual activity are a technological creation. Our sexuality is, in fact, technologically constructed and they represent a layer of sexual intrigue that ultimately leads to sexual intercourse and then reproductive evolution. But dolphins wear no cloths so the idea of posing for Playboy probably never occurred to them! Their sexuality is, no doubt, based on a different mode of sensory consciousness which may involve a sonic bodily impact for stimulated pleasure. Pregnant women have often reported feeling blasts of energy directed at their stomachs in the company of dolphins and considering the dolphins supersensitive ultra delicate skin, this evolution is indeed, I believe, most probable. It may even be that the sonic bodily impact of sound between mother and child are a vital occurrence for providing a foundation for lifelong emotional development. This may then translate into sexual communication between adult dolphins later in life. This is truly one of the great distinctions in the sexual consciousness of intelligent beings; because while human beings rely primarily on sight during sex, dolphins may rely heavily on the sonic impact upon their bodies generated from their own sonar blasts or sound communication. Imagine someone blasting a sound wave towards your body and you then receiving sexual pleasure from it.

The Dolphin PhD

In the evolution of such a huge and complex brain, tremendous and spontaneous communication has, no doubt, been developed involving a vast array of day to day communication for sustainment. And, no doubt, in the evolution of such communication tremendous organization has been established involving, perhaps, actual dolphin experts in various fields of dolphin sustainment by dolphins themselves. And because in an organizational development and because all dolphins are different, only the best should assume those particular analytical and communicational roles of which they have exceled. In their early evolution, sexual communication and the necessary communication of food gathering probably occurred first. But Later, as a greater evolution and awareness of the environment developed, many other aspects of environmental interpretation became available; geographical navigation, weather interpretation, and conscious predator and prey analysis, would, I'm sure, all be priority communication of dolphin society. But in the evolution of true dolphin intelligence an evolutionary network far beyond these communications may have been established; a communications network that has, spanned not only the entire globe for millions of years but also includes the great whales as well. There very well could be dolphins and whales who act as mediators between the great continents telling strange stories of far off adventures of other dolphins and whales on the other side of the planet. They may, in fact, have created a communicational directional linkage, with the knowledge that the Earth is round? Do they, in fact, wonder what might exist between the thousands upon thousands of miles of landmass between the great oceans? Perhaps they do; and dolphins have always been insatiably curious of everything, especially human beings; probably because we represent such a far out and completely different evolution yet at the same time share anatomical similarities with dolphins that are undeniable . Our precision cutlery of geometric angles and curvature is, no doubt, a whole new concept of intelligence to them. Geographic mapping of dolphin intelligence is, I believe at this point in dolphin evolution, a well studied and conceived art by dolphins themselves. And while within the consciousness of a relative motion and communicational symmetry they may have found the ability to receive sonic images of the geographic display of their environment indispensible. Dolphins have fantastic memories, and in the open ocean there may very well be dolphins who assume brilliant sonar mapping studies and communications; and it may very well be that dolphins, with perhaps finely tuned memories of sonic images, will assume these roles. But in dolphin analysis of geographic terrain, other perceptions of analysis may also become apparent. For instance sperm whales may be able to sonar up to 6 miles. If dolphins could pick up these sonar deflections as well it would provide additional information about their environment. This may aid in them in becoming more aware through sound and perhaps through motion currents of their geographic position. It may also allow them to comprehend and analyze sound channels in relation to the motions of the currents in the determining of their own oceanographic positions. Perhaps there are dolphins who are specialists in analyzing and memorizing the sound vibrations reflected from the deep ocean terrain as picked up from the network of the great whales. Specialized dolphins who are experts in deep ocean comprehension may be part of a network that has existed for perhaps a million years or more. There may even be dolphin scientists who analyze and catalog vast arrays of sound frequencies under the ocean for weather and climate studies. This would no doubt be an important role in dolphin evolution. However, it is likely in any evolution of organization that the best of talent will be brought forward to assume the tasks of which they may be best suited. And in the evolution of a dolphin network, these talents and perceptions may be divided into several dolphins. But communication of sonic image and sonar analysis would be crucial.

By now dolphins have clearly evolved a wonderful intelligence of play; a true sign of an intelligent being. Simple survival in the open ocean though, tough as it may be, simply does not, I believe, require a brain of that complexity. Dolphins have taken the conscious evolution of sound into an as yet uncharted direction of consciousness; a direction which may indeed involve new directions of storytelling and analysis of sonic concepts for the sophisticated exchange of highly evolved rhythmic pleasures.

In the actual consideration of intelligent beings in the Universe, human beings, having only just now reached our birth of technological construction in outer space, and are by no means knowledgeable of the evolution of cosmic intelligence to any significant degree. We have never encountered beings of sophisticated intelligence from outer space; at least not that we know of and certainly not on a global scale. However, in consideration of intelligent beings we do have an ongoing directional path to which we may turn; a path 4 1/2 billion years in the making; a path to which we may track the acceleration of evolution from a micro bacteria to conscious technological construction in outer space; and embedded in the rock and mud of nature is an imprinted record of its past evolutionary achievements. In its existence today lies a super spectrum of life forms spanning all levels of evolution from viroid development to conscious humanity. The fossil record of life on Earth exists in paged layers like a book. Generally speaking, the deeper you go the more ancient life forms you may find. And by tracking this directional evolution we can now determine characteristics we believe truly intelligent beings must have. Perhaps 200 million years ago a bizarre thing began to happen in our evolution. In the midst of a constant battle for survival against predators, many of which were ferocious dinosaurs, our offspring began to become more and more dependent upon their mothers at birth. Before, in the chaos, they were born ready to fend and fight for themselves to the death. Now they were becoming more and more helpless by the millennium. It seems to me that on at least a one dimensional plane of thought that it would be far more logical to have offspring become even more better suited to survive at birth then they were before; and indeed evolution does begin like this; where as thousands of offspring are born, ready to fend for themselves, in the hopes that one or two will survive to pass on their parents genes to the next generation. But Evolution had other ideas and it knew intelligence was the key to survival in an insane and chaotic environment. This meant that evolution could benefit in the development of beings who became more and more dependent at birth and as a result it became more beneficial to have just a few offspring and then spend a quality amount of time teaching them what to do and what not to do with emotional bonding as the magnet for continued communication and learning. This meant that now offspring began requiring more and more physical and emotional care from their parent beings to aide in their survival; It also meant that communication became a direct line to what was needed to evolve. Now the parent is engaged in a far greater protection of the young as well as passing on vital information for survival down through the generations. And unlike the hardwired instinctive communication of insect evolution, this new evolution of mammals from reptiles allowed a greater spontaneity of communication to better understand and react to situations based on their individual uniqueness. It was the survival of emotional exchange that would ultimately flourish.

Today we understand all of the most intelligent animals of the planet Earth to exhibit tremendous care in the raising of their offspring. The dependency of offspring at birth allows tremendous intimacy to grow between parent and offspring; and animals need successful communication between themselves in order to successfully learn their environment. It is the communication of dependency that provides the wonderful evolution of play to develop the complex knowledge needed to strategize for the benefit of passing on one's own genes with play as the highest learning mechanism possible. Today the ongoing evolution of dependency expands still outward and onward in the evolution of the Planet Earth. Today human babies are born utterly helpless and require constant care and attention in their first years of their lives. 200 million years ago our reptilian ancestors were born ready to hunt and fend for themselves; but it was the emotional exchange strategy that was the more beneficial.

Dolphins also have evolved into tremendous dependency of offspring as well. They are immensely sociable creatures relying on one another not only for survival but also for emotional communication. And like the human, play has evolved into a lifelong recreational enjoyment. Why, because 'play' is evolutions highest order of learning. This may be because 'play' is the ultimate fuel efficiency for maximum learning in this atomic dimension.

In scientific analysis of dolphin brain evolution, I believe, many scientists have misinterpreted the overall interpretation of what the dolphin brain is. They have, in many ways, attempted to evaluate dolphin intelligence based on land brained mammalian comparisons. Why because land based mammals constitute the overwhelming bulk of human knowledge of brain evolution. From mice to rats to dogs to monkeys to human beings, man has acquired much of his knowledge from land brain mammalian evolution. And while all of these animals may differ in a basic pattern of general brain evolution; dolphins have evolved in the water for the past 70 million years and most likely have evolved a completely unknown brain structure adapted for a completely different environment. However, 70 million years ago dolphins were primitive shrew like mammals on the land only then beginning an evolution into the water. Therefore they too may have possessed a primitive land based brain that then began an evolution into a new completely different environment. And, of course, 70 million years later it has most likely evolved a totally unknown brain structure. However, scientists know the basic structures of land based mammalian brains not only because they have a vast array of them to study in the lab but also because those land based brains are also more relatable. But the dolphin brain is most likely structured in a different direction of environmental evolution. And since dolphins communicate well beyond our range of audio perception and have no hands to manipulate objects, it is very hard for us to comprehend what may be going on in their brains thus making it difficult observe what may be happening socially. And it is here that we come to where we may have made a crucial error involving a lack of evolutionary insight on our part. Many scientists have claimed that while dolphins do contain an extensive fissure evolution of the frontal lobe cortex, which is crucial to the seat of reasoning and free association, they lack extensive development of other brain areas that they have recognized extensively in land mammals. In actuality, since the dolphin did begin as a land mammal and then evolved into the water for 70 million years, these land based brain functions should be degenerating or transforming into unrecognizable brain function because it is now engaged in an aquatic evolution. This means that now that it is water based it may no longer need them thus giving the illusion that it fails in comparison to land based mammals such as rabbits and even mice. Where in actuality, it no longer needs these land based functions because it has evolved a brain in another direction of cosmic environmental consciousness. In essence, men do not know what dolphins have evolved into and should not compare them to something completely different from dolphins. But in the future we shall, I believe, come to understand what a real intelligent consciousness in another direction of cosmic awareness is and hopefully gain a greater understanding of it for our coming birth into the galactic community.

The Digital Dolphin of Tomorrow

The dolphins and whales of the Planet Earth will be among these non-technological intelligences; and the dolphin in particular will be among the first to enter the digital realm. But for humanity, the journey towards the communication with another intelligence will not be simple. Artificial intelligence will have to be written for the transposition of knowledge. In the case of the Bottlenose dolphin, high pitched sounds ten times higher than a human can hear must be transposed into a visual math for humans to see and then comprehend. The process of communication with dolphin intelligence will, no doubt, be long and complex and will most definitely require a digital evolution. In fact, only in the digital age of computers can it be done. In the beginning we, as human beings will only observe as the dolphin learns to explore its new super toy of digital processation. This, essentially, will be an underwater recording studio, that at least in the beginning, will have huge foot long buttons for the dolphin to press with their snouts to manipulate sounds in everyway possible. It may sound ridiculous but this is necessary because we as human beings must be able to visually see everything that they do in order to understand them. They will be able to record, transmit, mix and be as creative as they want with sound all within this new realm of digital technology; and we as human beings will simply observe what they do as it is then translated into a visual math. Eventually we may learn enough to begin a primitive conversation with them on such topics as the melting of ice or something else of extreme simplicity. But in order to be able to do this we must then be able to transpose the sonic math that we can see back into the high pitched frequency sounds for the dolphin to hear and understand. Eventually we will begin to decipher the language of a new direction of consciousness. Ultimately what does this mean, you might ask? It means that we will discover new ways of comprehending the dynamic interactions of matter that may be geometric, mathematical, or evolutionary that human beings have never thought of. Dolphin intelligence may provide some of these new avenues of thought in a Universe of infinite complexity.

The great exchange of dolphin intelligence and 21st century digital intelligence has yet to take place. But when it does, it is here that human beings will truly understand the true power of dolphin intelligence in the Universe. And it is digital intelligence that will act as the referee between human and dolphin interaction. And though many have worked to decipher dolphin intelligence in the past, it is something that can only be accomplished in the digital age; because for a dolphin, to exchange information with a dolphin computer would require computer hardware and software that has yet to be invented. But if a computer were invented for the dolphin to input its own sonic 3D perceptions for its own unique comprehension then this would no doubt create a new science of interpretation for both dolphins and humans.

In the future dolphins may also receive and transmit information through the internet to other Dolphins around the world. Pretty crazy huh? Obviously it is human beings who would build such devices for the dolphin and this will probably be the first time human beings will ever build a device solely for use by another intelligence. For first time we will watch as another intelligence gives birth to its own digital information evolution and if we are smart we will start by just listening. Ultimately this uniting of dolphin intelligence with our own digital technology may become a magnitudinal super scope of emotional interchange in our super distant evolution. Both man and dolphin will experience the wonder of this emotional interchange as they step out of their little cradle in space and begin their long journey inward towards the center of the galaxy. But for now, at this present time, the exact second of our birth of technological construction in outer space occurring in the same second as our birth of artificial intelligence, we must begin to understand dolphin intelligence in the most 'out of the box' modes of thought ever imagined.

Eventually as dolphins become more advanced they too will experience their own mass communication birth in a digital context. Eventually we will see the first dolphin astronauts because we are now in the age where we know there's an entire ocean underneath the surface of Europa (a distant moon of Jupiter). And what better astronaut ambassador then a bottlenose dolphin to land on Europa, drill down into the ice and say 'hello'. That's why the emergence of the dolphin astronaut, however psychotic that it may sound, maybe crucial in the understanding of whatever intelligence resides beneath the ice of Europa.

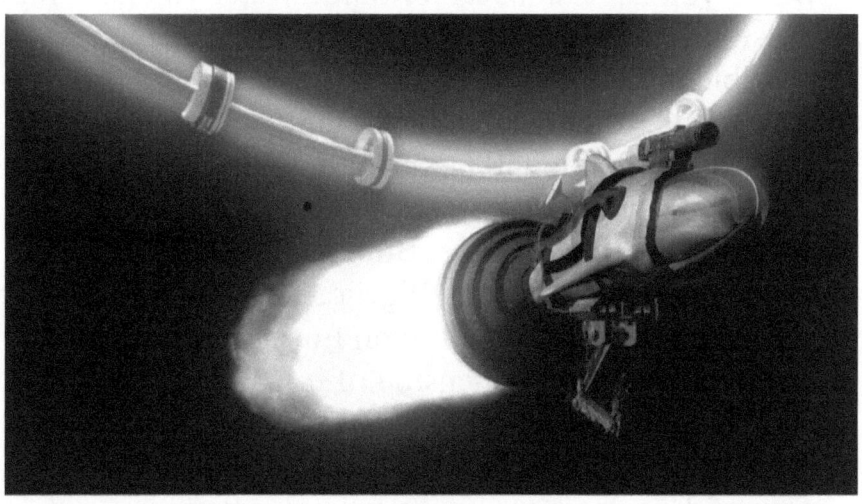

A dolphin astronaut of the future explores deep space from the safety of it's centrifuge.

The Dolphin/Europa Connection

Because of this and because the Universe may be populated with non-technological intelligent beings, the recent discovery by the Galileo spacecraft of cracked ice on the surface of Europa has indicated the possible presence of liquid water below the surface. In 1996 the Galileo Spacecraft photographed the surface of Europa, the fourth moon of Jupiter. It showed the surface to be populated with an amazing array of cracks which is usually indicative of a liquid ocean below. This makes sense because as Jupiter constantly tugs on Europa it distorts it from a normal round shape to a more oblong shape which means tremendous heat is built up at the core causing the melting of the ice throughout certain levels of the Europa. This, of course, gives you an ocean; and where you have an ocean of liquid water you have the possibility of life; and after 3 or 4 billion years it is possible intelligent life could arise. Bizarre as this may sound, if there is, in fact, intelligent life on Europa, a dolphin will be the only thing that will be able to understand it correctly. Why?; because it will be an aquatic intelligence. But equally important is if there is not conscious intelligence but, perhaps, an eco-habitat of microbial life on Europa then once again dolphin comprehension of it would prove invaluable. Even if no life exists on Europa at all, dolphin comprehension would again represent a powerful geologic tool in aquatic planetary understanding. It certainly isn't guaranteed that life exists on Europa; and while life could probably survive; there is still the question of whether it could arise in the first place at all. I guess it's possible there could be some impediment for life not existing on Europa but considering the extreme divergence of which life that has been discovered on the Earth, it is very possible that in a water environment and during the course of more than 4 billions years, life would arise; especially taking in to account that Europa has probably been bombarded by asteroid and cometary impacts, many of which may contain life generating amino acids. Whether or not life exists there or not, one thing should be understood; dolphin intelligence is something that is completely outside the realm of human experience. Those who undertake the goal of comprehending it must always be wary of the Flintstones. (example: projecting modern human qualities on people who lived 1000's of years ago.) While this might be funny to watch it wouldn't teach us much about how cro-magnon people actually lived. In fact, it would actually distort data. Meaning we must always be aware to not project our own human qualities into our understanding of it. Those who only look for similarities between ourselves and dolphin intelligence will fail to fully understand the direction of this super consciousness.

*Above: Three dolphins arrive at Europa by their own
free will aboard their massive centrifuge.*

Dolphins and the Center of the Galaxy

Above: Three sperm whales monitor their squid farm constructed in a deep space centrifuge.

In an amazing array of underwater vocalizations indecipherable to present homo sapiens technological abilities, the great brains of dolphins and whales exchange consciousness in an awesome display of unconditional care and affection, loving one another in an intricate pattern of emotional display that symbolizes the essence of freedom and love in the open ocean. On the other hand; who knows what the hell they think about. They seem to exist in the extremes of love, sex, and violence; and this may very well be true everywhere in the Universe. In fact Evolution demands it! If it were not for violence, death, and mayhem how could evolution function? I mean that's one of the Universes main hobbies!

Anyway, like I was saying before, the great brains of dolphins and whales roam the open oceans of the Planet Earth in an intricate understanding and comprehension of their environment. But some of their future belongs in outer space. Why?; because the dolphin brain is a mind of fantastic analysis and there are alien intelligences near the center of the galaxy that would like to talk to it!

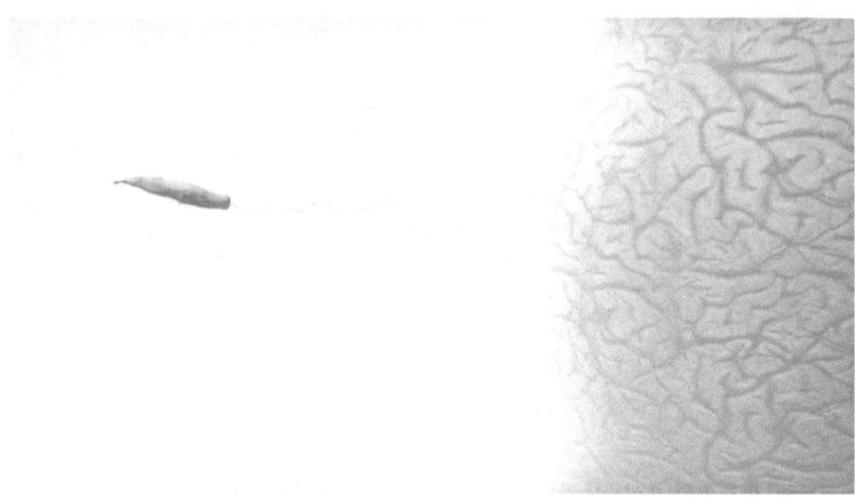

A sperm whale of the future makes its way towards the center of the galaxy via homo sapiens transport to encounter an ancient massive brain of whose intelligence may be a million years beyond our own.

Its perceptions and comprehension of matter are, no doubt, unique in the Universe. And what, exactly, are they saying, you might ask? What do dolphins discuss with their massive brains? Change; change in their environment. Let me explain! The ocean is an environment in constant flux. Everything is in motion including the ground. Even the rock formations undergo change at a variable rate. To live in this environment and engage in a discussion about it, one would definitely have to discuss its constant change. And sound would be the ideal medium to perceive it with. Amongst the many things that could be perceived and discussed with sound could be the ocean currents, fish populations and potential breeding, shark behavior as well as food chain analysis, and general change of any matter that might exist in the ocean. By understanding the sonic math of all these and other things combined, one could reach bewildering conclusions about one's own environment. This communication of change in an environment in constant flux may represent a huge aspect of humpback whale song which many believe are used as mating tactics for breeding. Conceivably these songs could represent a type of summation or newscast that allows conveyance of an overall view of the ocean environment. A well done assessment could impress a female enough for some serious sex. That's what I'd use it for! But because humpback whales use longer traveling, lower frequency sound to perceive and communicate with, they may engage in a much broader, more general, sense of the ocean environment. It may be the smaller toothed whales like dolphins and porpoises, who use higher frequency sonar to perceive with, that engage in a more microscopic analysis of the environment.

Both dolphins and the great whales, however, may exchange information with one another or perhaps combine their knowledge of the environment to reach overall conclusions of the current ocean state. There are, potentially, a great many things to discuss in the ocean and no doubt the dolphin mind is aware of things I couldn't possibly think of. Its comprehension of shark behavior is no doubt an intricate study of behavioral instinct. And while it has mastered a comprehension of its universe in a consciously aware state, it has also demonstrated a spectacular display of unconditional compassion. It is an intelligent being, and like all intelligent beings it loves to feel the awe of emotion; to socialize in the ecstasy of the conscious exchange of thought and perception. The understanding of dolphin intelligence is crucial to humanities understanding of the Universe. I realize it doesn't seem like it now; but as we delve deeper into this particular neurological science it will become more and more important to our information processation. Why you might ask?; it is because human beings do not as yet understand three dimensional sonar comprehension and the kind of data it provides. But let us also remember that it is the dolphin's birth of digital/technological comprehension, as built by human beings, that will now morph their existence into the information highway that is created technologically by human beings.

Why Europa could be a land planet

(No I'm not on any medication)

One of the reasons that it would be nice to know whether there is life on Europa is not just because it would be nice to know if there is life on Europa, but because it would give us a better idea of how much life may exist on all these ice worlds that are being discovered. It's not really possible to gain a proper census of what distribution of life there is in the Milky Way Galaxy until we are able to take a reasonably bad census on how many of these ice worlds may be harboring life per solar system. There could be multiple ice worlds harboring life deep below the surface just in our own solar system; so any attempted tally of life distribution would no doubt be wrong. So right now we on the Planet Earth are just trying to get to the point where we are just plain ole wrong as opposed to catastrophically wrong on what this number might be. There could be far more of these types of worlds then perhaps, any other type of world that may harbor life. To be clear; this type of world is where internal heating caused by gravitational perturbation has caused layers of ice to melt into water forming underground oceans possibly allowing the evolution of life to happen. Assuming that life aboard these worlds is prolific, then the question begs. Is it all aquatic? Or could it be that a multitude of different kinds of life evolution can take place. These ice worlds have really messed up my life because I was so sure that I had the entire Universe figured out up to this point. To make matters worse I am about to commit the ultimate sin that can be committed by any free thinking cosmic theorist who prides himself on always thinking outside of the box; and that is to take a normally freaked out world that is doing nothing but minding its own business and apply Earth like standards to it because guess what? I'm from Earth. But dammit, Europa really could be a land planet. Consider, for example, a simple predator/prey relationship that could exist in the deep ocean on Europa with an estimated 60 miles of ice above. Imagine for instance, the prey, while being chased, hits the bottom of that ice and, for the next 500 million years of evolution, begins to tunnel its way through that ice for the purposes of safely creating an elaborate array of ice cities essentially making a world like Europa a land based world. In this sense it would be a duel environmental world with an ocean beginning some sixty miles below the surface and a huge network of tunnel configurations carved into the floating ice above it. Ancient microbial life could theoretically alter the atmospheric chemistry to an oxygen concentrate much the way it happened here on the Earth billions of years ago. And yes, beings could actually be walking around up there and on a low gravity world like Europa they could theoretically be huge. It is questionable as to whether the necessary metals would exist there so far from the sun for any sort of technological evolution as we know it to exist on Earth; but at this point we need to be open to different kinds evolution whether they be technological or aquatic or something in between; and we have to remember that a Europan ocean that exists below all that ice has been sitting there for 4 billion years or so. It has been bombarded by countless cometary and asteroid impacts courtesy of Jupiter. Many of those impacts may have injected crucial amino acids; the building blocks of life; into the substrata of Europa for genesis into God knows what creatures that have now been hanging out on Europa just waiting for some technological species like humanity to come along and attempt extraction or in Dick Cheney's case; drill for oil. Because as we all know; even though they have been sitting there for four billion years and everything has been fine; they need to be rescued. Remember any bold and adventurous Europan that decides to reach for the stars by tunneling its way through all that ice to the surface is going to get a real rude awakening when he finds himself face to face with the mighty Jupiter and all that wonderful radiation that could fry an egg in about a second. I guess this would be their version of Icarus who flew to close to the Sun. Perhaps the bold really have made their way to the surface of Europa and actually glimpsed the beautiful colors of Jupiter only to duck back down below quickly and survive to tell the story… possibly getting laid as a result of their bravery. In short; don't mess with Big Daddy Jupiter! Remember billions of years ago when Mars tried to be friends with Jupiter … when Mars just said "Hi Jupiter, My name is Mars. Will you be my friend?" and Jupiter said "Yes my little red friend. Come closer. Closer! Closer. Moooooo Ha ha ha ha!

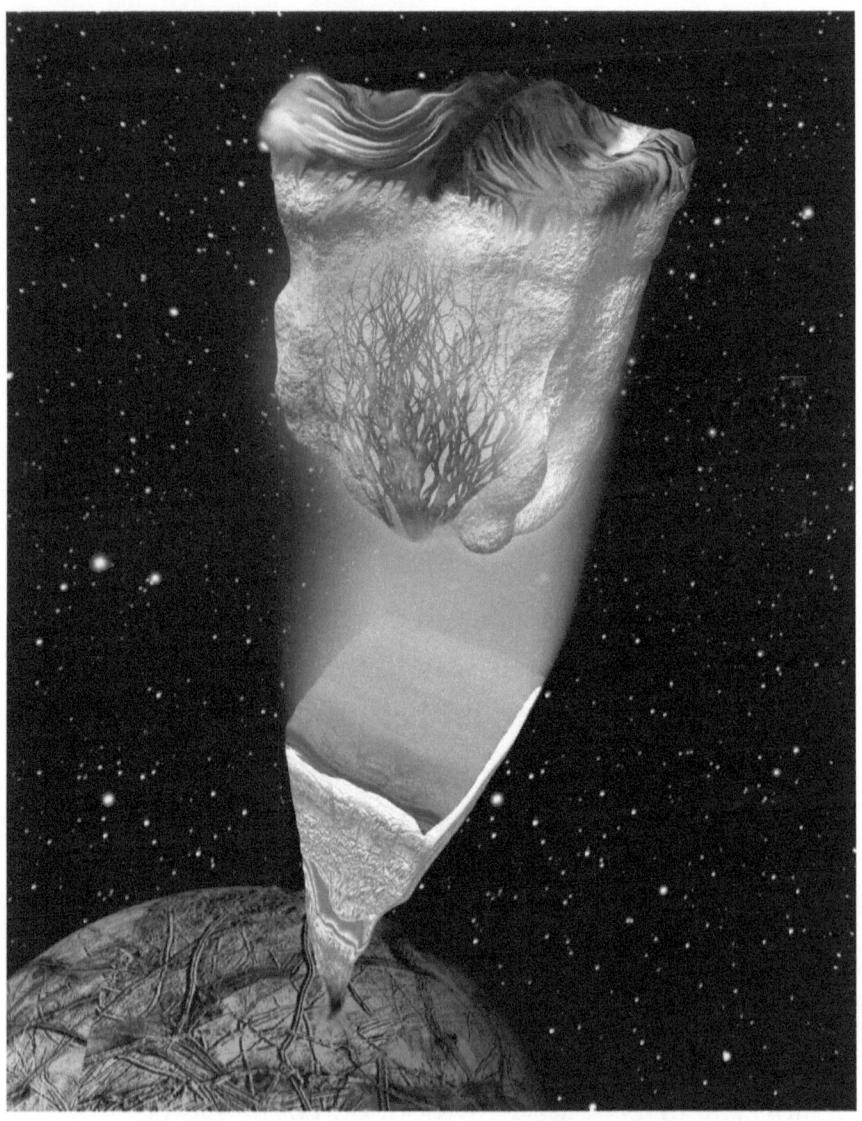

Previous: As we can see from this slice of Europa, the green network of tunnels would extend from the bottom of the ice upward. After a billion years it could be extensive enough to allow unimaginable alien civilizations to develop.

Binary Star Bizarre

As we make our way out into the Universe we may very well find that there are classifications of alien life that we can expect to find in certain environmental conditions. And it is important to understand that the environment determines the direction of evolution that a species will take. This is probably never been more true than in the binary star system. This is where the really freaking shit exists in the galaxy. These are the true psycho solar systems of this atomic dimension. It's all those seasonal changes from being affected by the gravity of two stars. Hibernation, multiformationality and evolution in slow motion are the trillion year long fads. As a matter of fact it is estimated that some of these low mass stars may last as long as ten trillion years giving plenty of time for whatever long range hibernation species to adapt. If ever there were a place beyond human imagination then these planets of binary star systems would most certainly qualify. What might we find you asked? I'm seeing a base life form like a sponge or maybe the atmosphere itself as protective coating for the planet that orbits two stars or at least has to deal with the temperature extremes of two stars. It probably wouldn't be a higher order intelligence and might be akin to the foundation of the food chain similar to what plants do on the Earth. This would be a bit different though in the sense that it coats or encompasses the planet with the ability to predict and respond to temperature changes that would be happening on a fairly regular basis. The higher order intelligences would then have a parasitic relationship with it and be protected from the outside influx of heat and radiation from a probable red dwarf star. The higher life forms may also have the ability to withstand vast temperature changes themselves. The dynamic of such an ecosystem could get quite complex with thousands or millions of different species all integrated into the foundation of a base life form that provides, not only protection from the orbiting stars above, but may also act as a food source to be farmed by the higher order intelligences that are dependent on it. There may be hibernational territorial disputes between species factions as well as monetary incentive at the expense of this protection shield by some species of the planet. Sound Familiar? The complexities of such an ecosystems could be vast. Imagine trying to explain the concept of a reverse mortgage to a dolphin. There a concepts involving life form interactions in this atomic dimension that might not be comprehensible unless you are actually integrated into the society. On the other hand the base life form that I just mentioned, might be intelligent after all and have to ability to undergo intelligent chemical change to deal with extreme weather conditions that might only occur every 10,000 years or so. Perhaps a brain is required for this type of organism to survive on one of these crazy seasonal insanity planets. Or maybe the base life form, which as I said, is the foundation for planetary inhabitation, operates similar to the way ants operate on this planet. It is made up of trillions of life forms all dedicated to the whole as an end result. They are spread out around the planet constantly gathering intel on the constantly shifting weather patterns and communicating it to other members for eventual group analysis. Perhaps there are conflicts between those who want what's best for the species as a whole and those who have alternative agendas similar to what we call greed on the planet Earth. Or perhaps what they do on one of these psycho planets is so far out that I can't even begin to put my finger on it.

Imagining aliens by imagining their environments

If I had to imagine what alien life might look like I wouldn't approach the process by drawing some crazy looking creature with 20 eyes, 8 hands and 6 faces. Instead, I would look at what the environment was. It is the environment that determines, not only what the species looks like, but what the species actually is and does. Different environmental arenas will produce vastly different civilizations of life forms and many of those life forms that have evolved in one environmental arena may very well bleed over into other environmental arenas as well.

On the Earth we have many different types of life form civilizations. And it could be said that we have at least three environmental domains of which civilizations have evolved; the land, the water, and the air and with significant bleed over from one domain into another. And these environments can, in themselves, being broken down into different environmental arenas to produce vastly different civilizations. Elephant intelligence which is more like the sonic consciousness of the whale has developed on the land. Birds occupy the land too but can spend and enormous amount of time in the air. Ocean sponges tend to conglomerate together to form a whole being. Seals, sea lions and walruses occupy both the sea and the land. Caterpillars transform into butterflies having emerged from embryonic cocoons to then take flight and become a duel environmental civilization. Almost all of these species mentioned here would be considered 'multiple environmental' to some degree or another. What's also important to understand here is that civilizations, no matter how small, have different evolutionary foundations in their basic structure. For instance, Bees seem to exist in a strange sexual civilization where as all individual members are components of a hive colony dedicated to the fertility of the Queen. If this basic foundation of colony were to survive and undergo a 300 million year evolutionary synthesis into conscious self awareness with a hive mentally at its base, it would be vastly different than anything we as humans have ever imagined. Dolphins would represent yet another genre of foundational civilization type. Here we have an aquatic sonic consciousness that no doubt has already developed into a sophistication perhaps as complex or even more complex than our own. And because the medium of communication is so radically different than our own, it extends out in a direction of which the details contained within it would be mindboggling to comprehend. It's also possible that it could contain significant elements of a hive consciousness as indicated by their extreme social behavior.

Almost all civilizations contain some element of a hive mentality. If you look at human civilization you will see the collective consciousness at work in a great many areas. People form groups along with other people that they don't even know exist. Whether you are rooting for a baseball team or sharing political views with other humans you are in a hive colony dedicated to the benefit of that cause. The 'vote' in a democratic society is a hive function.

Humans, of course, occupy all three domains with the foundation of our civilization taking place on land and because of this we represent the manipulation of matter into tool for the benefit and demise of virtually all life on Earth with the possible exception of the deep ocean. There is a collective brain aspect to humanity in our civilization but we also exist strongly as individuals separate from other people. Dolphin intelligence can also exist as a separate entity. But hive mentalities exist in degrees and something like an ant civilization which seems to be a kind of collective intelligence whereas all individuals seem to be only components of a central brain, has a much greater degree of hive mentality. The question being, of course, what does a super consciousness, that has a large magnitudinal hive foundation to its evolutionary upbringing, look like? Are they self aware or does the collective actions of the whole become a self aware entity? Now combine this with ideas of information intake that are foreign to human understanding and you will get a primitive idea of how vastly different and varied the Universe can be in terms of alien differentiation. Here on the Earth we seem to have a multitude of different foundations for colony differentiation. And while the environment of the species does determine what the species is, there are other aspects that go along with it. Basic predator/prey relationships play a huge role in sculpting out what a species must become. Also, the medium of information intake and communication in relation to predator/prey relationships as well can all combine to form vastly different types of colony consciousness.

It's important to understand that civilization traits like these, if allowed to evolve as a foundation for super intelligent civilizations, would probably evolve characteristics so alien that we as human beings might not even be able to comprehend what it is that we are looking at. As said before it is difficult to ascertain an inventory of what we might find in the galaxy without actually experiencing the many different kinds of planetary formation. Even in our own Solar System the idea of having an understanding of the many possible environments that can form for the evolution of life to happen is rapidly becoming more ridiculous the more that we learn about it. Thirty years ago nobody had any idea that all these ice worlds, that are harboring liquid oceans, even existed. And each one that we find may very well have a different evolutionary foundation that its indigenous life forms may have undergone.

Of all the strange evolutions that may exist in the Universe those of which inhabit the planets of binary star systems may be the most complex and bizarre; at least relative to what we have here on the Earth. A binary star system is, of course, a double star system with two suns of various types orbiting one another. It appears that approximately half of all star systems in the Milky Way are, in fact, binary systems. They come in a variety of star types and configurations involving brown or red dwarf suns orbiting anything from each other to pulsars, to white dwarfs to red giants and even black holes; all these combined with virtually unlimited orbital configurations can produce temperature variations that, for whatever life forms may exist there, to evolve in. On Earth, of course, we have only one sun of which the Earth orbits on a twenty three degree axis creating the four seasons of which much of the life on the Earth has evolved to thrive in. But in a binary star system we have two suns generating heat in orbit around one another. This means that whatever planets that are in orbit around one or both of these suns that are hosting indigenous life forms will experience spectacularly complex seasonal changes. The reason is because the orbit of two suns is a vastly more complex thermo dynamic than just a single G type star like the Sun. This means that whatever planet is in orbit around one or both of these suns will undergo a far more differentiated temperature flux in what very well may be a far more complex orbital configuration. This, in turn, means that whatever life forms that may inhabit these types of worlds must adapt to potentially wild temperature increases and decreases in their quest to survive. At times the planet may reach boiling temperatures only to cool drastically into subzero temperatures in very short amounts of time. This all depends on a wide variety of circumstances involving planet rotation, tilt and orbital configuration of the binary suns in relation to each other. It also means that whatever life forms that exist in orbit around these suns must be able to predict and then react to the constantly changing seasonal changes that are always happening. It may be that there are life forms there that can exist only on one side of the planet; constantly moving against the planets rotation to stay within a habitable temperature zone. Or there may be a goldilocks ring running perpendicular with the planets rotation that life forms must exist in because both sides of the planet are uninhabitable as it rotates. It may be thousands of years before you experience the same season twice. Or you may experience time spans of which the planet may undergo a freeze situation and as a result everybody has to hibernate. Sometimes the surface of the planet might not be habitable; at least not in the domain of which you are the most comfortable. Deep planning by life forms, both cognitively and genetically, may be necessary to survive. If you know that a hot summer is coming in 200 years then you may have to start growing and planning now for future generations to survive deep into the future long after you are gone. This means entire civilizations may have to move to more habitable zones of the planet over a period of decades in order to survive the coming routine orbital apocalypse. Sometimes maybe only some of the life forms may find the planet habitable; this may mean hibernation for deep extended periods of time; thousands of years perhaps, while some other species of life of which you have never seen, has awakened and is now thriving in that environment until it then goes back into a deep hibernation because of the ever changing orbital seasons. Sometimes it may be necessary for a life form to hibernate for a few thousand years only to wake up and find the place trashed by the jerk who was out of hibernation 500 years ago. Sometimes it may be necessary for a life form to change their design or form to adapt to a constantly changing environment. Try to imagine a caterpillar transforming into a butterfly times a thousand. In fact, being a shape shifter as presented on Star Trek might not be a bad idea. But instead of just transforming ones shapes into something else it might be necessary to undergo a complex cellular change as well. Whatever the case it has the potential to be a vast and completely different ecological evolution unlike anything we have ever seen here on Earth. The fact is that if you are a planet orbiting a pair of stars, of which the orbital configuration could be a vast combination involving stellar masses at various distances, you are going to experience a wide variety of seasonal changes: In other words, instead of just 4 seasons like we have on Earth there may be a 1000 or 10's of thousands of seasons.

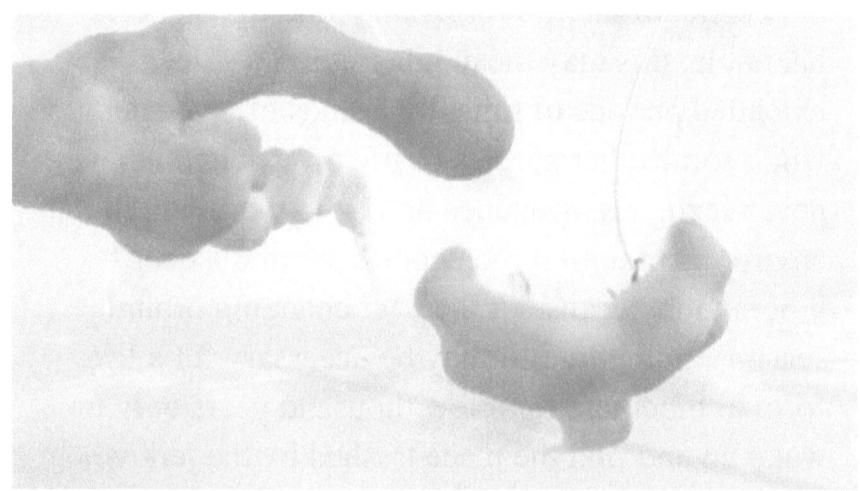

Evolution on a planet of a binary star system may produce a higher degree of multi formational or shape shifter life because of the multitude of seasonal changes that would happen because the planet orbits two suns instead of just one.

The Third Gravity

Other types of worlds also may exist that have never been thought of. Again we don't know what we don't know. Theoretically, if an alien species could blow chunks off of a neutron star, which is basically the leftover remnants of a star that exploded, strange little worlds of tremendous gravity could be produced. Remember, an ordinary teaspoon of neutron star matter which may weigh as much as Manhattan or even the state of Ohio could supply a decent gravitational pull if separated and harnessed.

Theoretically, if you can get a plasma beam or maybe an antimatter plasma beam to blow a chunk off of a neutron star and that chunk actually maintained its density you could theoretically pave around it and have whole world only a mile or so across with approximately one Earth gravity weight. The inner core would only be about the size of a basketball and the process might look something like this;

Young naïve pulsar minding it's own business receives visitor from outside it's orbital domain;

Inside the invader centrifuge lies a nuclear weapon gatling gun that could literally cut a hole in a planet if it needed to.

Aboard the 500 foot gun structure lies an alien hillbilly like architecture bent on extracting a piece of

pulsar matter for the purposes of constructing a small mile wide gravity world upon which the resident alien could then vacation.

Our little alien then prepares his weapon and takes aim.

He then ignites it at the surface of our innocent pulsar, which by the way, never meant to hurt anyone.

As the super hot nuclear antimatter plasma beam bombards the pulsar it causes dense pulsar matter to eject from the surface.

The ejected matter then begins it's journey through

space

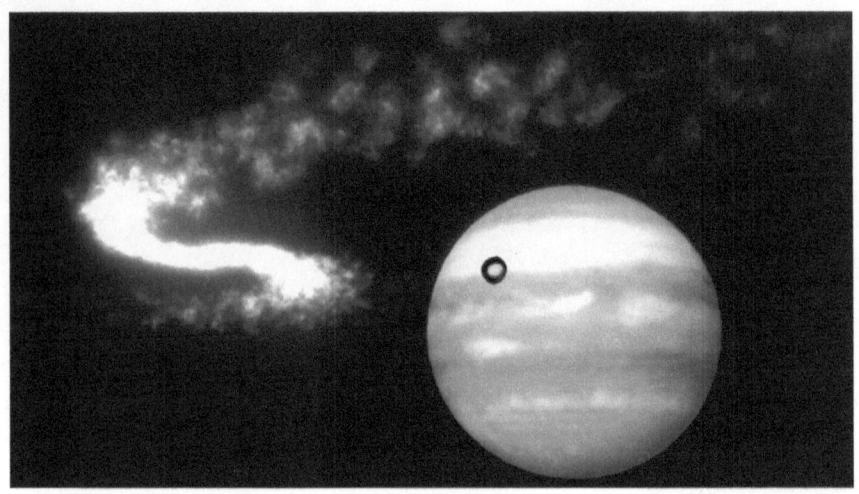

of which it finds itself captured in the orbit of a large hydrogen world of which then enables the ejected matter to condense into a nice round basketball sized sphere.

From this point on construction around the super hot sphere begins. This involves constructing a mile wide

spherical architecture for habitation with the now condensed pulsar matter that will provide suitable gravitation.

The final result is a mile wide world of which a person can live and farm in a suitable gravity.

After that the occupants can live happily ever after complete with love, great food, and spectacular sex;

and if you are really desperate you can actually see the sex in the youtube video 'The Third Gravity'.

Ok; it probably wouldn't look like this; and I admit it's an absolutely a crazy idea mainly because the heat and radiation being emitted from this basketball sized ball sphere at the center would fry about anything that it approached. However, if you could divert the heat away somehow, it might be feasible. There would, of course, be other massive problems as well such as; is it even possible to create a weapon that would blast off chunks of pulsar matter far enough away to be usable? And would that matter still maintain its density and be gravitationally suitable? And how long would it take to re-condense if you even had a large jovian planet in exactly the right position to allow it to condense? On top of that; what would you do with all that heat being emitted from the center of your mile wide world? Personally, I wouldn't go near the dam thing; as a matter of fact just forget I brought the whole thing up.

Disclaimer: Do not try this at home.

THE GODS ARE CHASING THEIR ANIMATIONS INTO INFINITY

Or astrophysics through animated cartoons
(As if that actually helps)

But they will never catch them. It is doubtful they care. They're having too much fun. They love the exploration of the physics and evolution that awaits them in their distant future; so do human beings for that matter. Every second they are evolving in huge explosions of biological creations, yet they are completely at home in their constantly evolving environment. For us it would be like waking up in the morning with a normal DVD player and by the time you got through with breakfast it had evolved into a multi-functional virtual reality hologram environment of simulated physics and not to your surprise. This is what alien intelligences undergo on a daily basis. They are forever becoming children again and again into infinity; and they're having a blast; so why stop? These aren't just regular children either; they are super comic insanity starbabies of incredibly advanced evolutions! But why would they be chasing their animations into infinity?; Probably because the way a technological species draws its animations may actually predict its own next stage in evolution; and it will happen to us too. In fact, it is already happening. Our earthly and beyond earthly desires are will always be hinted at in our superheroes and cartoon adventures. Superman would be good example. Why?; because Superman possesses traits that are extremely desirable in our real life environment that is always trying to kill us. Human beings understand these talents and as a result, pursue these talents through the natural evolution of technology. But this concept of pop culture evolution that, in effect, binds reality and fantasy together exists across the entire entertainment medium. As a matter of fact, almost all of the entertainment we create actually hints at our own next stage in evolution. We are headed into infinity and Superman is waiting for us in our distant evolution. We are evolving and in a billion years we will be something else beyond our present imaginations. Superman is a dimensional suggestion of what we might become; an aura of biological physics accommodation; the super ability of biological physics adaption to an environment. This means that if a species is to survive and prosper they must become Superman in terms of evolution. Evolution, in fact, demands it. But before that happens physically, it will be suggested in our entertainment to ourselves. Beings of unimaginably complex outer universal physics will, through natural selection, evolve a biological physics adaption that is relative to their entertainment environment. The concept of Superman is one of our first earthly contacts with this distant super evolution of our future because someday, like Superman, we will fly with advanced propulsion, have super strength through genetic manipulation, and be able to see through walls because perversion is one of the great necessities of invention. Theoretically though, his ability to have super strength, fly, and see through women's clothing could make him the ultimate indestructible terminator peeping tom. Superman is not alone either. There is, of course, a multitude of pop culture heroes that all have different talents and powers that we will ultimately adopt in our real world evolution of the future. In terms of say, Bugs Bunny, the basic body structure alone is in line with current and past evolutionary developments. But, of course, it depends on who is drawing the character as to what will happen in terms of suggesting the future. The late Warner Brothers animator and director Chuck Jones may have been the most adept at generating images and motion that may have suggested what lies in the future for humanity. As an animator, Jones was superb beyond description. He always drew his own poses for the key frames of his animations therefore he controlled the poses and motion of his characters; and he was ultra self aware of the physicality he was drawing. As a matter of fact, the poses and animated motions that were drawn by Chuck Jones may represent the highest level of motion and pose ever created by human evolution. The characters themselves (Bugs Bunny, Daffy Duck, Wile E. Coyote, amongst others) are usually represented in general human form somewhat beyond the physicality of the modern homo sapiens human. In fact, they suggest elements of the next stage of human evolution when placed in front of modern man on one of those evolution 'march of progression' graphics that we have all experienced.

As man evolved from a four legged quadruped to an upright two legged simian his posture slowly approached the completely straight vertebrate posture of Bugs Bunny and many other cartoon characters. Usually such characters are drawn with a slightly more inwardly curved lumbar representing a step ahead of where humanities skeletal matrix appears to be headed. Let us now look at the insanity of the Looney Tunes universe to see what lies ahead for the future of human evolution.

Comic Insanity: Comic insanity is a main staple of animation cartoons and, of course, the comic chaos of an animated cartoon also appears to be present in real world mammalian evolution as well. The effects of the comedy upon the brain have always been beneficial and appear to increase with increased intelligence and because cartoons are only imagined and then drawn then it makes sense that they should present the greatest form of human exaggerated communication. It also makes sense that what we exaggerate for entertainment is in direct relation to what we may need or desire in the future.

Altered Physics: As one could probably guess; the physics of the cartoon universe strays somewhat from the elements of reality that we have grown so accustomed to; but our own altered physics are just around the corner as ever increasing digital processors are paving the way for us to eventually inhabit the digital physics world of our choice. So eventually if you want to walk off a cliff and not actually fall until we realize it; our dream may soon come true. At the rate we are going, Google Earth should eventually engulf our entire reality plus the Earth in general in about a thousand years. From there your digital avatar body will be able to experience whatever physics it desires.

Planatary Terraformation: Animated cartoons have long been known to alter entire landscapes and even entire solar systems. In reality, we are probably on this course as begun by the invention of farming thousands of years ago … and from the construction of The Great Wall of China to the Panama Canal, to the Kansai Airport (which is still sinking, by the way) this evolution of terra transformation of the Earth will, no doubt, increase in magnitude indefinitely, perhaps to the eventuality of the construction of a Dyson Sphere that surrounds the entire Solar System as imagined by present day engineers. In addition to that, the eventual evolution of digital physics landscapes will allow outrageous creations of environments for human habitation. Ultimately, the convergence of reality terraformation and digital creation may create environments far beyond what can even be comprehended by the human mind.

Increased Intelligence Through Living with Humans: In terms of the greatest relative mind expansion, I would say the dog probably benefits the most. After all they are intuitively smart and yet live with a superior intelligence of which allows a huge range of mind expansion to occur. Remember, the key word is 'relative'. Human beings only have each other. There is no higher intelligence of which to grow from. We don't relate to dolphin intelligence yet. But the domesticated dog is subjected to concepts in human society that it would never encounter in the wild. It has to grow at some level or levels within its well developed mind. It may very well be that they are amongst the top one percent of one percent of all intelligences when compared with all life in the galaxy.

In animated cartoons, animals, of course, are almost always presented at a level of human intelligence. In the world of Looney Tunes animals seem to exist in a quasi neutral state of their natural habitat and a state of equal intelligence with the humans that they seem to do battle with. Remember that time Bugs Bunny had to do battle with an entire city to force them to re route a massive highway on ramp in order to preserve his rabbit hole. Spoiler: The city eventually blinked first. In the real world of hyper relative evolution such bizarre scenarios may actually happen, relatively speaking that is … especially in a world where digital intelligences will be evolving at spectacularly fast intervals. Already we have experienced significant evolutionary speed increases in just in our own lifetimes and in the future, A.I. or even human intelligences may start off at a particular rate and yet, evolve at a multifold rate in a relatively short amount of time. Moreover, domesticated animals may, themselves, undergo tremendous evolutionary intelligence increases as a result of direct manipulation of their brains from human or possibly dolphin intelligence. This could allow the reality of a Bugs Bunny to actual kick our asses someday. Call me a traitor but I rooting for the Rabbit.

Accelerated Evolution through Societies and Cultures: Naturally with the ability to transform entire landscapes, and directly cause intelligence increase, it should be no surprise that entire cultures should evolve right before our eyes. Bugs seems to have been key figure in the discovery of America in 1492 yet also seems to exist in modern day times as well as every other time period in between. Humans will do this too as we become more and more immersed in our digitally projected world of which we have already begun to explore. If you lived from 1500 to 1600, not a whole lot would change in terms of innovation. Nowadays, I could wake up tomorrow and find that I am behind the times in terms of the latest digital culture shock that struck, seemingly overnight.

Coordination beyond human ability: Where else could we experience the ultimate comprehension of speed but in an animated cartoon. It is the simplest form of communication that doesn't require a miracle of the laws of physics to convey. The cartoons of Chuck Jones who directed many of the great Warner Brother/Looney Tunes cartoons may be the highest form of speed and motion coordination ever achieved by human evolution. Jones always drew his own poses and key frames for his cartoons requiring his in betweener animators to comply with his drawings. The result was spectacular exaggeration performed with coordination that will only be achieved a million years from now when we become gods in the real world of actual physics.

Speed: Jones also did the Road Runner Cartoons. The desert settings were a perfect place for the two wiring characters of the Road Runner and Wile E. Coyote to do battle in a way that would explore the wonders of caricatured physics. The timing of the Road Runner when it shoots off and causes the road to rise up from the loss of air pressure below is truly a remarkable thing to experience. It is timed to the exact frame, of course, for maximum effect. Only a god could move like this. Human beings, also, have chased higher and higher levels of speed since the invention of the wheel.

Intelligence: Cartoon characters are also able to think

quickly and complexly. They have to since there is only a limited amount of time in a given performance. Still the way they think can influence the way we think in the real world; and, of course, being able to analyze and think quickly is always at the forefront of evolutionary survival.

Globalization: Whatever a cartoon character needs seems to exist immediately off screen. As human technology progresses via the internet, along with jet propulsion, the ability to physically possess whatever we desire not only increases in terms of it's variety but also in its speed of delivering. Thousands of years ago horse travel was the fastest way to deliver goods and in yet another evolution that is taking place in our lifetimes, 3D computers are now spitting goods and services of use right in the comfort of our own homes.

.

 Cuteness: Yes, cartoon characters are almost always cutie pies of one sort or another. But cuteness has always had an enlarged role in the survival of the fittest; particularly in the age of mammals. The parental attractions towards the mysterious equations of cutie pie ness have allowed mammals to thrive for hundreds of millions of years. They are a prime motivation for communication in environments that kill relentlessly those who lack crucial information about those environments. As we have seen, 'selection' in terms of who survives and who doesn't often pits cuteness against cunning with cuteness winning out because empathy and love are often the overwhelming forces that determine the end result. Personal case in point; when I invited my white kitty to come in my house and live with me it was because she couldn't compete a much larger and more experienced bully tom cat that was dominating the food sources. The tom cat may have been stronger but it was no match for the power of snowbitties cutie pie ness, therefore the weaker cutie was awarded a new home with a 24 hour food source while the stronger tom cat was forced to endure a less fortunate fate outdoors. But the power of the cute and lovable are not just silly little aspects of the society we live in; they now determine what direction human evolution will take. If we prefer to draw cute cartoon creatures because it ignites euphoric empathy in our consciousness then we will also ensure their survival in the real world over creatures that may have been dominate in the wild.

This is because cartoon characters are almost always drawn in a way that suggests what we as human beings desire most in lovability; big eyes, enlarged head, feet and hands and, of course, comic chaos. In short; the reality that we live in can never catch up to the desired exaggerations of our imagination. We will always be chasing our animations into infinity. What does it all mean? It means that ten thousand years from now we will all be walking like Bugs Bunny and talking like Tweety Pie. Don't fight it.

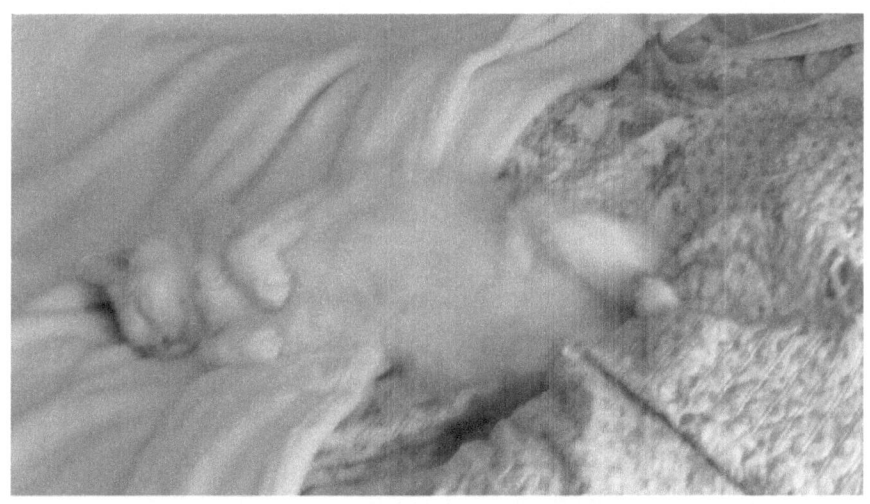

In this atomic dimension we have cuteness powered by stellar evolution and in terms of evolution it could be argued that the adorable will kick the ass of about anything else that gets in it's way.

Interestingly enough is also the concept of neoteny. Neoteny is a strange phenomenon in that a baby chimpanzee will share many of the characteristics of an adult human as opposed to an adult chimpanzee that will possess more ape like features. This is an anthropological concept that many scientists say is ongoing in terms of human evolution and it seems to be connected in strange ways to the ideas of exaggerated cuteness as presented in our imaginations.

The aspect of cartoon characters being presented in desirable exaggerated forms usually seems to involve every known characteristic of human endearment. Only in terms of endowment do cartoon characters seem to be lacking but I got a feeling that if some of these little cartoon characters were to unzip their little cartoon suits they would be packing some serious sausage underneath the hood.

Flintstones; meet the Flintstones

There are many aspects to the animations we draw. Some, as mentioned above, suggest aspects we may develop to avoid thermonuclear destruction. Others may be combinations of our many different epochs in history. And some may be common to many intelligent species in the Universe. Let us begin by imagining a technological civilization discovering its own creative engines to explore the fantasies of it newly born conscious imagination. Naturally, of course, being more familiar with itself than anything else it tends to generate fantasies involving different versions of itself and its environment more than anything else. And, of course, since it is at least as intelligent as we, it is reasonable to assume it has a sense of humor evolved out of its conscious awareness. All of this is more or less general in the evolution of technological intelligence, I believe. Now let us assume this intelligence has adequate knowledge of its fossil evolution leading up to its present evolution. It knows how its deep ancestors lived and is familiar with the transition between its present and past evolutionary states. One day one of their own hits upon an interesting comic thought; what if their previous distant ancestors, in all their primitive ways lived similar to the ways we do today with technology based upon history. Gee, what a funny thought; but somehow logical. Upon further consideration of this strange comic inspiration our being may find it far more funny to envision their past versions of their own present technology rather than just their modern technology in straight form. And with this they have discovered what may occur quite often in the Universe. They have created the Flintstones; a perfectly natural conscious consideration in the exploration of one's own mind. It is not freak or special consideration created specifically by the human race; it occurs everywhere in the Universe and, of course, it's not called the Flintstones, as if I actually had to tell you that. This concept will most likely eventually be stumbled on by someone in whatever part of the galaxy or dimension they may inhabit. It's all about seeing one's self through the eyes of others. You've no doubt heard of BAT-MAN. But have you ever heard of BAT-POODLE? Or maybe a story about a cat that has the ability to turn into a sperm whale in times of crisis? No, because nobody can relate to it unless it involves humanity or some aspect of humanity. The Flintstones represent a good, logical example of this type of thought entertainment that no doubt is a common theme in the Universe; and that, quite simply, is when an intelligence species combines its present state of existence with aspects of its past state of existence for the probable purposes of entertainment. In a way, Star Wars does the same thing. It is about history too. It combines the future of technology with historical reference.

It sounds simplistic and innocent enough yet you have to realize eventually it does begin to take on responsibilities of raising our children. And while I realize many of you think that this may be a bit silly it is indeed probable and possible many of our present day engineers were directly influenced and inspired by the Flintstones as children. After all, children don't just grow up and then suddenly have desires to become inventors or designers right out of the blue. They develop and relate concepts they see all through their childhoods. Inside the mind of a child, the wheels are always turning, applying everything they see to develop a conscious understanding of their Universe; and the Flintstones offer a spectacular display of creative ideas and suggestions to consider. So now suddenly the Flintstones, a mass communication creation, became a major force in the evolution of humanity affecting God knows what, in the course of human evolution. Now I realize many of you may think this is a bit crazy. I realize by now many of you think I'm a bit crazy. But seriously, if a kid watches the Flintstones at an early age and then ultimately, because of it, grows up to become an engineer who develops a crucial piece of technology that then goes on the change the world for better or worse, well, that's the reality of the world we live in! Wouldn't it be wild if some major world altering evolutionary event took place and then unbeknownst to the world, the key to it happening lie hidden in some Flintstones cartoon that aired some 40 years before. There are untold numbers of little events like this that alter our evolution but are never known by human beings. The Flintstones, I believe, are indeed, only one type of universal occurrence that may occur throughout the Universe. But there are, I believe, a great many others.

"If it ain't broke, don't fix it!

Yes, any technological intelligence has probably made this statement because certainly at some point one of its beings will try to tinker with something unnecessarily and someone else will note that the device is not broken and that it does not need to be fixed. It is probably not an uncommon occurrence in the Universe to make that statement. There are probably a great many repeating themes that we as human beings may think are unique to ourselves but, in fact, are commonalities throughout this atomic dimension. How about "Stop n Go". Here's a general concept of a necessity to perform something quickly and rapidly so one can move on with one's life. It is an idea that is initiating a quick "in and out" to receive something with a name using two of the most basic syntactic gestures that can probably ever be created and applying them directly to the prime purpose of the reason for the whole thing. "Stop n Goes" have no doubt, littered this galaxy for millions of years and they are not about to stop.

TOMORROW NEVER DIES

Welcome to my new super number:

500 billion electrons could fit on the head of a pin.

The speed of light is 186,000 miles per second (around the Earth 7 times in a second)

The Milky Way galaxy is 100,000 light years long and about a 1000 light years in depth.

THE NUMBER IS: a '1' with the number of zeros after it equal to the amount of electrons that could you stuff into the volume of the Milky Way Galaxy? That's a big number. Written out in numeric form it would be '10 to the i power' whereas the 10 stands for 10 and the 'i' stands for insanity.

So the questions is; does the Universe repeat in cycles of similarity over eons of time or does it not repeat in cycles of similarity over eons of time? There are several questions here; and to be honest with you, about the only thing that I can do is ask the question; the answers may very well be tucked away in the hidden mysteries of infinity.

First; do events repeat again throughout infinity exactly. My answer would be 'no'. Mark Twain said it best when he said "nature does not repeat itself; but it does rhyme". In order for the events to repeat exactly then all of the events throughout infinity would also have to repeat at the same time since any one of the these 'so called repeated Earths' would extend out and connect to the rest of the Universe of which created it. This would mean that the entire Universe would have to repeat exactly. So I would say 'no' on that one; but what about another version of the Earth complete with basic homo sapiens evolution occurring again with different events. Let's use the James Bond adventures as an example for kicks and giggles. Let's say the Sun forms then the Earth forms, homo sapiens evolve, the English language happens, and then the Sean Connery version of James Bond occurs along with all of his adventures and then the Sun burns out 4 billion years later. Then '10 to the i power' years later another Sun forms along with another Earth; homo sapiens evolution happens yet again as well as the formation of the English language, but now the Roger Moore version of James Bond occurs along with all of his adventures before the sun then burns out 4 billion years later and the process repeats. Most likely if these events are to repeat in terms of harmonic evolution then all of the atomic particles involved will have to be in different positions relative to each other. If all of the atomic particles were in the same exact positions relative to each other then the Sean Connery adventures would repeat with only the Sean Connery adventures. So with two different James Bonds we have two different relative positions of the atomic makeup happening. Yet the two James Bond occurrences may still be related to one another directly in terms of harmonic evolution even though they have occurred a 10 to the i power of years apart. They would be an example of harmonic evolution on a massive time scale. And harmonic evolution is, in fact, embedded in the law physics of everyday occurrence. It is a repeating phenomenon that occurs in octaves as demonstrated in the real world physics of musical notation. But how close to a repeat does it get? Remember, only with infinity can we expect these kinds of phenomenon to repeat in some fashion. So the question is; are there other Earths out there that are also called 'Earth'?; and do the events repeat close enough to the point where I could walk through one of these 'repeated Earths' and feel at home. So lets get a little more specific; what if Sean Connery's 'James Bond' in the film 'Thunderball' occurred as said above. (*Lets assume its real! A real stretch of the imagination, I know*) Sean Connery, of course, played "James Bond" in 'Thunderball' which was about the evil Spectre gaining procession of two nuclear warheads for the purposes of blackmailing the U.N. for a cash advance. So let us now imagine that once again the galaxies have formed, stellar evolution happens, newly born planets form along with human evolution but now, 10 to the i years later; we have Sean Connery's "James Bond" repeating in the same type situation in 'Never Say Never Again' *(a remake of* ' Thunderball*)* which once again involves the evil Spectre gaining possession of two nuclear warheads for the purposes of blackmailing the U.N. for a cash advance. This time we have the same James Bond and same story but different events. In other words; very similar person engaged in similar events. In this scenario the events will repeat themselves over and over again into infinity but each time it changes slightly. The big question then becomes; will these events repeat themselves so far into future that the Universe will allow them to repeat almost exactly but not quite exactly; and how many different ways can it repeat in some fashion? In the above examples we two different James Bonds and then we have the same James Bond repeating the same story but not exactly. Remember also, for it to repeat exactly then the entire infinite Universe must repeat exactly because it is unlikely that we would have a slightly different stellar evolution creating an exact replication of events in terms of James Bond adventures. In other words everything is slightly different or everything is exactly the same and most likely everything would be slightly different. This would be another example harmonic evolution happening in different way. It is important to remember that these things can only happen if the

Universe is infinite because only then can we deal with these kinds of crazy numbers. What important to understand is that we are really looking at the many different ways that harmonic evolution can take place. Instead of just the two examples mentioned above there may be an infinite amount of ways that events could repeat. I chose the James Bond adventures to present these ideas because movies are a great way to present different event phenomenon. But we all know that events really don't happen in real life the way that they happen in the movies.

For example, consider that the events that occur in the Universe have what I call 'event expansion limitation'. Event expansion limitation refers to the idea that reality will only occur to a limited degree of exaggerate. In other words, in the James Bond movie 'Octopussy' no laws of physics are broken as far as I know yet these types of events don't happen in the real world. In other words, the reality of events as they happen to us everyday won't occur to any degree of consistency the way they do in a James Bond movie even though the James Bond movie broke no laws of physics. Another way of understanding it would be to imagine building a soccer ball that surrounds the entire solar system with the word 'Nike' engraved on it. Is it possible?; of course, but has anybody ever done it in terms of infinity?; probably not. This concepts applies significantly when attempting to decipher what it is that infinity can produce in terms of events repeating themselves over eons of time. My personal belief is that what we are experiencing now will never happen again exactly but I might change my mind 5 minutes from now. The reason is that infinity science is a difficult thing to contemplate probably because my brain is finite but infinity is, well, infinite. But it does suggest a hauntingly strange universal phenomenon. The "James Bond" scenarios mentioned above could repeat in terms of harmonic evolution in terms of either infinite time or infinite space of which is my way of stating that reality could repeat in terms of harmonic evolution in terms of either infinite time or infinite space. So some point, in terms of events we have to acknowledge that crazy movie events probably don't happen despite the fact that those events may not violate the laws of physics. The physics of what can actually happen is constrained by the exact construction of the particle dynamics that make us up so instead it is more likely that the events in something like 'Moonraker' do violate the laws of physics but do it in a way that we don't completely understand. And by the way, I just remembered that the character 'Jaws' in 'Moonraker' bit a two inch steel cable in half; so yes; shocker; 'Moonraker' does violate the laws of known physics.

It begs the question; is the nature of the atom able to extend into infinity without ever repeating? And if it does then it brings up a host of bizarre questions. For instance; what if I just draw a straight line out into infinite space? Would the Universe repeat eventually with another 'Me' and how close to this 'Me' does it get? What if I imagine infinity repeating into the past or the future? Do I repeat there as well and how exact of a repeat is it? Believe it or not, this question has actual practical applications for our everyday lives. For instance, let's say that we all do repeat almost exactly to the point where we couldn't tell the difference between one Paul Vancleve and another Paul Vancleve. The question is; what if my consciousness just bounces around entering and leaving an infinity of physical matter of Paul Vancleve bodies spread throughout the infinite space and time continuum that are all so alike that we could never tell the difference between one and the other. If I have a bad and day and make a bad decision then my consciousness may land in a Paul Vancleve body where the events are ready to flow negatively at me to let me have what I deserve for making an idiot decision of some kind. The ultimate irony of this theoretical phenomenon is that we, as individuals, may be intimately closer to other people that may be physically distant at ridiculous distances (remember the 10 to the I number).

Infinity thinks it's cool just because it's infinite

It is the complete Universe, completed though infinity that exists all at once; it is all the mass and everything that can ever happen existing throughout infinity bundled into a single thing. Everything that can ever happen, happens all at once. It is the natural existence of the naturally occurring mathematical fractal essence which has always existed since the beginning of time which never happened; and this is the creation of matter. It is all the physics of mathematical essence which must exist at once. The time when you were 5 years old and shit your pants at school *(I'll admit it if you will)* will live forever. And there is probably a cosmic record of it too. It will never die! It is has happened, happening and will always exist into Eternity. Why is it we don't experience all of time like this? It is because the physics of the Universe propagate in only one direction and since our brains are made of matter that must obey the laws of physics, our brains can only experience time in the direction that physics propagates. The whole Universe could be being rewound by God and played backwards like a movie reel and we would still perceive time as moving forward. Even if the Universe exists all at once and there is no movement of time we will always still perceive time as moving forward because our brains are composed of physics that will always propagate in one direction. And it will always be happening into forever. As a matter of fact, if you were to break matter down forever into smaller and smaller bits all you will ever find is attraction/repulsion forces of space and time. Just the same if you were able to look at this whole universe as an atom in a plate of spaghetti then the combined gravity of everything would appear as a particle. Matter as we think of it, is way overrated and can't exist in the Universe. It is fractal essence. (ie: gravity, electromagnetism); and, of course, velocity being the super engine of the Universe, can only occur in the form of mass. All of the Universe is all one thing and if you want to separate it then it must be in terms of relativities. The natural occurrence of mathematical fractal essence in the form of Evolution is the great miracle of us. But...

Why do objects attract?

Gravity! But what is attraction really? What makes one object attract another in terms of space and time. Is it those gluon things scientists have proposed that makes particles stick to one another. Doubtful! Gravity is a holy thing! It's not just some particle you can assign a tube of superglue to and then trot off on your merry way! It's conceptual foundations represent the very basic construction of the naturally occurring atomic Universe.

Gravity is probably as close as one can get to God, at least in terms of theoretical physics. It may very well be the product of other forces of attraction and repulsion that emanate from the earliest beginnings of infinity which never happened. In other words never yet forever! In the most real sense gravity may be a type of time in its actual physical construction. But why do objects attract one another? The answer is very difficult to conceive in a three dimensional universe, but I'll try!

First we must understand that all the matter in this Universe wants to occupy the same space at the same time in the same way that, say, the number 3 as a concept always exists in the same metaphysical space as the number 3. Face it, 3 is 3 which is 3. You can't have 2 different 3's. If you have the idea of 3 then wherever the idea of 3 is, it will be the same thing precisely as all the other 3's. Matter is the same way. Matter basically says to itself "look we are all matter so we should all exist in the same place. But the nature of matter has limits to it that don't allow this to happen. Think of it this way; because two protons are the same kind of mass they will want to exist within the same space and time. That's generally speaking! But the particle dynamics of the proton will only allow so much interaction before the nuclear repulsion that occurs within protons prevents protons from coming together *(fusion withstanding)*. Matter is also made of things that absolutely have balance in how much will exist in a given amount of space and time. So you have attraction at some levels such as gravity and you have repulsion at other levels such as repulsive nuclear forces. Somehow matter got separated from itself and just wants to bring all know matter together in the same point in space and time. But there is something about mass that prevents it all from coming together in a single place. In this dimension it's called fusion. If it weren't for the fusion that takes place in the interiors of stars then theoretically all matter would just come together into an infinitely large black hole. The Universe is a mighty strange place. But to understand why particles are able to attract one another is to understand that in essence they are the same thing and are both equally as big as the Universe itself therefore they should exist precisely in the same place except for the fact that naturally occurring repulsive force math will only allow so much proton interaction in terms of space and time.

Lets look at it another way. To truly understand matter then you must understand the true concept of essence. Let's say you are the number 3. Where are you, 3? The answer is that you are everywhere; and you are always sitting between the numbers 2 and 4, relatively speaking; nothing can ever change that. But because you are 3 then you are everywhere; everywhere relative to both 2 and 4. You are as big as the Universe but the Universe is infinite therefore 2 and 4 are also as big and always next to you. But to truly understand this you have to understand what it means to be the essence of something. Because you are the number 3 then you are mathematically accurate and connected to all the rest of the mathematically accurate Universe of unchanging fractal math. You are as big as the Universe but the Universe is infinitely big. In terms of the actual 'you'; you are only your consciousness which is made of matter that is mathematically perfect in its accuracy across infinity. Your consciousness is as big as the Universe but so is all of the matter relative to it. And you are your consciousness in true essence. It is the only thing that you will ever experience and you are trapped in a Universe where the idea of space is an illusion as to what matter really is. Space, in fact, is the only way that our brains, which are made of matter conforming to the laws of physics, can perceive the Universe.

No person or scientist in the world can state by even the remotest proof that things do occur randomly. That's just as wrong as saying that things do absolutely occur in order. But it could be that beings millions or even billions of years beyond myself can see this with the greatest of ease. They, no doubt, have reached a consciousness where they can easily see that all events can only occur one way if given their mathematical positions in space and time. There is no chaos whatsoever!. Think about it like this. If asteroid "A" is traveling at 520 mph in a known direction and asteroid "B" is traveling at 260 mph in a known direction we conclude they will collide at point "C" provided that we are aware of all other outside forces in a mathematical way that could interfere. As long as we are aware of all the mathematical speeds, positions and directions of everything in space and time just like the Universe is, we can determine absolutely what these asteroids will do. Nothing can prevent this because we are aware of everything's position and all of the other aspects of them such as speed and direction. What is the point? The point is, is that if we know all the speeds and directions and positions of everything in the Universe we can determine exactly what will happen. But most importantly of all, we can clearly see only one possible future can happen. All possible forces are known in every detail. Using this logic I have successfully deduced that the Universe can only occur in one set of events. Of course it's a big theory and I'm open to suggestion but I'm not really sure there is anyway around it. Of course, we

don't actually see it this way because there is no way
for us to know all the mathematical positions of
everything in the Universe. Yet we can deduce it may
be happening. Most people can understand this but
most people don't understand this, but it is not because
they are stupid, it is because it is kind of a depressing
thought. Therefore they reject it. The idea that we
have only one future no matter what we do is a little
disconcerting. However while many people will agree
with it in theory, many have suggested that while
inanimate events may occur in only one possible
outcome, the conscious decisions of living beings are
unpredictable. Yet like the inanimate world, the world
of our brains is also nothing more than particles
bouncing off one another. All are susceptible to the
same laws of physics. Let us now imagine Bob. Its
late and he's tired, but he needs milk for in the
morning . All he wants to do is go to sleep. But he
knows he should pick up milk tonight or he will hate
himself in the morning. What should he do? To
imagine this, we must understand that in the decision
making process, the brain at any frozen moment in
time is, of course, in a particular state and that Bob
has many different feelings and emotions ready to
intermix to produce a decision. However we must also
understand that if we were able to view all of his
emotions and their interactions with one another, we
once again could determine exactly what decisions
would take place down to an infinite level. We would
easily be able to see what emotions would be
dominate or grow dominate and foresee exactly the
state of his brain afterwards. If we could freeze Bob's
brain in time at any given moment and had total

understanding of his brain and the events surrounding it at that same moment we could calculate that only one possible thing could and then would happen. It is my belief that the Universe does indeed occur in only one future. I know it's not a popular theory, but as time and evolution move forward, future generations will easily deduce this as almost certain fact. But does it occur in an exact event direction for the purpose of evolution?

Whether it does or doesn't, one thing I do believe is, no matter how intelligent or conscious we become into infinity we will never be able to foresee what comes next. Somehow the Universe and infinity will always be ahead of us. It's like an unwritten law!

The faster dimensions ahead of us probably extend into infinity in the direction of increased velocity and by the same token the slower dimensions below us probably extend back into infinity. And while infinity may be infinity, how infinities are relative to one another may chart a new direction of universal understanding. Infinity is a strange thing. Strange questions occur when thinking about it. For instance, Let's say we are both gods and I ask you what was the biggest creature that ever lived in the history of the Universe? And you say 'Oh well, that's easy; It was creature X that lived on 'Planet Whatever' 150000000000000000000000000000000000000000000 000000000000000000000000000000000000000000000 000000000000000000000000000000000000000000000 000000000 years ago. It sounds like a simple question. Yet in thinking about it, it gets weird. Remember, we are talking about the infinity of time here. It doesn't seem right to say that maybe one being was the largest and that no other being into infinity was larger. I mean, let's face it; we are talking about all of infinity here which means all three dimensions of space into infinity; both directions of time meaning infinity past and infinity future and the infinite spectrum of atomic substance of both slower and faster atomic dimensions into infinity, as well as whatever infinities that you can find in the directions of nuclear funnelization. I can't think of any more but for all I know there could be an infinite amount of infinity types as well. By the way, feel free to combine infinities if you feel like. In terms of determining what the largest creature that ever lived in the Universe was; it seems to me that whatever being you find that you think is the largest ever, infinity will produce one even larger into infinity. But if you have infinity to deal with then conceivably you should always be able to find a life form larger than the next which means eventually you would have a being bigger than a galaxy. But that probably doesn't happen. What I think happens is there must be some point at which life forms can become no larger; which means, I guess, that you must always be able to find a larger creature that is bigger only by a size measurable by the infinite break down of the size of an electron or some microscopic amount; or in other words; infinite break down into larger and larger sizes but never surpassing the length of an electron; or any atomic particle for that matter. Infinity always does this and it drives me crazy. If only it weren't so infinite. But it is weird! And it is a question that has no answer because there may be properties about infinity that are beyond human imagination. This would be related to 'event expansion limitation' in that hidden within the particle dynamics of the atom are designs that prevent outrageous exaggeration of atomic matter events.

Are we happening elsewhere

This is one of those questions where a little acid never hurts. Let us now consider the insane. What if I draw a straight line in any particular direction with infinity to spend. Would I eventually happen upon something in disguisable from this present Earth environment. How close would it get. And if it got close, remember I still have infinity to get even closer. So the question is then, is there a version of this Earth where all the atoms involved are in slightly different positions. The problem with this is then as we expand out from the Earth then we find that the Sun and moon and everything else must theoretically be similar which means that the entire Universe would have to repeat itself every so often. Remember at the end of the day, the Universe is basically 2 + 2 equals 4. Whatever is mathematically possible will happen and nothing else exists. Even the concept of non existence cannot be imagined by conscious intelligences like ourselves. How could it. You would have to be a non existent unconscious entity to conceive it which, of course, make no sense. It really comes down to how much diversity is possible in the Universe and in particular, this atomic dimension.

Finding the Millennium Falcon

Let's think about the Millennium Falcon for a second, from Star Wars. It's that round saucer shaped hotrod pirate starship that was seen at the destruction of three death stars; not that I'm implying anything. The question is; if I go out and pick up a round naturally formed rock from my back yard then I can make the argument that it vaguely resembles the Falcon. But surely to God there has to be another naturally formed rock that looks even more like the Falcon somewhere on the Earth?; maybe it a bit flatter like a saucer; and then, of course, somewhere else there has to be another naturally formed rock that maybe looks even more like a saucer with, by chance, the protruding emergence of what looks like the cockpit on the right side. You get the idea, right? Now applying all those infinities I mentioned above we can surely see that there must be another rock that formed naturally that looks even more like the Millennium Falcon. The question is; how far does it go? Can I eventually find a rock in the same shape as the Millennium Falcon complete with hollowed out cockpit?; with maybe a couple of chairs thrown in to boot along with Han and Chewie at the controls? My only answer is that the resemblances must at some point break down into still more accurate looking naturally formed Millennium Falcons yet do not exceed a certain accuracy; once again 'event expansion limitation'. In other words infinity is being infinite yet never surpassing a certain point. It means Infinity finally ends … at fractal infinity which means it goes on forever yet never surpasses the point at which a naturally formed rock starts to rival a finally sculpted Millennium Falcon done by a person. A God might know more about this but for now you are stuck with me. And I suspect that no matter how intelligent you get (maybe a trillion years in advance of human intelligence) you will still not be able to comprehend infinity! Which is perhaps its greatest and most healthiest feature. Perhaps this is a true and defining feature of God; INFINITY; it does truly defy consciousness, period.

Will the collision of the Milky Way Galaxy and the M31 Galaxy in Andromeda yield a new kind of galactic bio dynamic

So far the Universe appears to be forever changing. From the time of the funnelization of our birth matter into this dimension, the distribution and atomic complexity of matter has been constantly changing. It took several billions of years for matter to evolve to a complexity for consciousness to happen. So why should we expect it to stop anytime soon. Soon, 4 billion years or so into the future, the Milky Way Galaxy will collide with the M31 Galaxy in the constellation of Andromeda. The collision of these two galaxies will create new and vast complexities of matter that will generate chemistry and compounds in quantities that will, in turn, generate new forms of civilizations that are not at this time, possible. The reason is that stellar evolution, amongst other things, such as pulsars and various other collisions, will have changed the complexity and distribution of matter to allow different forms of life to evolve that could not exist now. Scientists are fond of saying that when the collision takes place that the two galaxies will pass right through one another because of the vast distances between the stars. This is true but long term gravitational interactions between the stars will take their toll as well as the interactions of all of the gas and dust between the stars. It may very well be that the evolution of life in this Universe does get complex, not just in terms of species evolution, but in the complexity of the creation of life. In terms of the collision, believe it or not, it is actually a distant possibility that someone alive today could see it. How you might ask, well according to Carl Sagan, if we were accelerate at 1 G we could navigate the observable Universe in about 56 years ship time. We would just need a continuously burning fusion engine of which plans are already on the Drawing board. But for now we have our own issues to deal with at home with this current technological mass extinction on Earth of which I discuss in more detail in 'Trump and the Asteroid that killed the Dinosaurs'.

Does any lifeform species survive into infinity

This is by far my most cosmic question on a personal level. I will probably never know the answer in my lifetime. In fact, it may be that no intelligence ever really knows it. It seems like such a waste of all that time to not have any life forms survive indefinitely into infinity. Could it be that the laws of physics allow for spectacular technological innovations trillions of years in advancement yet no intelligence ever survives long enough ever develop them. Imagine the possibilities of a species that could evolve cognitively and technologically a trillion squared years into the future. There is some evidence that the Universe wants life forms to survive indefinitely. The relationship between the threat of thermonuclear destruction and our emergence into outer space appear to coincide together and we seem to be surviving the relationship only barely ahead of the curve. But the Universe may happen in a way that potential thermonuclear destruction is a crucial element in the painful evolution of a species. The keyword, of course, is 'potential'. It may not be that the Universe is trying to kill us. It may be that the only way the Universe can evolve super conscious intelligence is to bring it to the edge of death. Maybe it has no intent of killing us. Maybe the Universe is simply driving us to an indefinite evolution and mass destruction and the only way to get there is heaven and hell tying the knot. But just because Humanity may make it off of the Earth and safely into the cuddling arms of deep space where nothing can go wrong, doesn't mean that the Universe doesn't have other wonderful things awaiting us to ignite further the evolution of consciousness. Take for instance, black holes in space. Eventually we may reach the great black holes of the Milky Way Galaxy and prepare to fall into them to experience the perceived riches that may exist on the other side. It may all look great on paper but there may be a catch! While the math may show that humans can enter and then survive these other perceived dimensions on the other side of the black hole, it also may be reveal that no being that enters into these dimensional transformations may be able to declare its survival to its counterparts that it left behind. Imagine having tapped into the last of whatever is possible in this atomic matter technologically and thirsting for what may come next only to find that science and physics does not allow for the communication that everything is cool on the other side. We could, theoretically, propel the entire human/AI/dolphin population into the black hole under the calculations that everything should be fine on the other side yet never knowing that nobody ever survived and as a result all of civilization went extinct because somebody missed placed a decimal point in their calculations of black hole entry. This might be something that we might want to get a second opinion on. Maybe even a third.

If, however, life forms do go on to survive in some manner forever in order that the Universe can experience all that is possible then certainly the technological constructers (Humanity) of the Universe would most likely be employed to propel the non-technologicals (dolphin intelligence) off of the planet Earth before the Sun engulfs the Earth some 4 billion years from now. So the apparent hope and ability seem to be there.

Is humanity a television show and are we being viewed by any couch potato gods

If we come to survive the insanities of our societies

it might be nice to know if somebody recorded it for the purposes of entertainment. I'm tired of putting make up on everyday not really knowing whether I'm being watched or not. To make matters worse, it may not be a case of some other intelligence observing us but that we are being observed yet nobody gives a dam. Alien intelligences have lives too and may not necessarily want to be bothered every time some little mote of dust achieves a mass communication birth. However there may come a point in human evolution where the library of our lives actually becomes our lives. Right now if you want to learn something about humanity you might reference a data base or library of some kind. But the future may behold that all that we do may be also become recorded data in the same instance. In fact, we may already exist as pure data. This atomic dimension may be a library. It also may be a farm. In fact, it may be a farm, seeded by some cosmic god farmer, that happens in a library to be referenced, provided, of course, that whatever alien or god that attempts to look us up can provide the proper documentation. Whether any intelligence has ever actually checked us out for reference would be interesting to know. But a god might see things differently. It may see the all of mass communication birth as a whole entity or as a part of something else that we are completely unaware of. We truly do not know what we do not know and whether or not we are descended from a natural process of the Universe or we are generated as data in a Playstation 5000 simulation is debatable. If a god farmer did throw bunch of ordinary matter into the wind then it may have what I call 'quantum cowboys' to help control a proper outcome. This would be an intelligence beyond our comprehension that has seeded this particular atomic substance space with both the matter and the quantum cowboys to round it up and keep order. The God Farmer may only want control as to what kind of life may result from their seeding just like a regular Iowa farmer might. But let us understand that he would not be a person or being of singular proportion. A God Farmer would be a vast intelligence consisting of every intelligence concept that we have ever conceived of from neurological to digital to algorithmic to whatever we saw on Star Trek to anything else I might have missed and then some.

In short whatever concept that you have ever encountered involving intelligence then combine them all together and then magnify it by 10 trillion in terms of complexity. As a matter of fact, we better make it 100 trillion. On the other hand the Universe could be just the rawest of wild wild west of indifferent matter at it's most natural and barbaric state. Or maybe our God Farmer has seeded us with a bit of both. But most likely the Universe wants to feel the pleasure of consciousness as a zenith and that track of properly farmed play will find a lighted path by the construction of the hydrogen atom. I mean why would the Universe be interested in anything else other that feeling and experiencing a good laugh and maybe a little sex. It is the atomic horny that we live in, so we might as well enjoy it. It is that track of mounting comprehensional play within our consciousness that will lead to the highest and most desired state of the particle dynamics of this atomic dimension. Translation: That's us. We are atomic before we are anything else and the mysteries of the mind and consciousness are deeply connected to that atomic matrix in ways not understood. There may be benefits in the way we treat and experience consciousness in terms of play energy as a super fuel of conscious evolution. It, play energy, may be the final solution of the Universe or at least of this atomic dimension. Those who can engage in the highest levels of comprehensional play may be the ultimate in the survival of the fittest. And as we learn more of the wonders of the algorithmic play phenomenon of the brain, this atomic dimension may deliberately stay ahead of us with new and never before generated event scenarios to fuel the evolution of consciousness as much as possible. Often this will mean tragedy to generate evolution. After all this is an evolution dimension and no doubt higher intelligences have adopted an algorithmic approach to the comprehension of play upon the brain in order to adapt to this atomic dimension of which always happens in terms of pure evolution. The competition itself for higher learning will no doubt lead to the indoctrination of all members of a society, alien or not, into a comprehensional play cognitive that will engulf all of a society. This is what the Universe wants. We are the Universe and when we feel the pleasures of super play then the Universe does too. It may very well be the singular event to allow the ascension of existence beyond the threat of thermonuclear war. But these two things, algorithmic play and the threat of thermonuclear destruction, may coexist and even collate with one another for the maximum ascension of consciousness. No pain no gain is an evolutionary statement and wherever we go or whatever we become in the depths of outer space, somehow the threat of complete annihilation will always be present or in the process of occurring even over a number of centuries. The codes of the Universe require love and madness to create each other.

What is the relationship of humans and Earth long term or in other words 'What is it that we don't know that we don't know.

The answer is "I don't know". But human long term relationships with the battered planet Earth most certainly will happen even if visiting the Earth of the future means wearing a spacesuit. Humans have always been willing to occupy a vast array of uncomfortable environments and after 10 years of Donald Trump, I'm ready for life on Titan without a space suit. And unlike Mars, the Ileen Wournos of the Solar System, you can actually get out and start walking around on Titan. Oh sure, it wouldn't be fun, but you could do for a few seconds. As a matter of fact, Titan is the only moon or solar body that could actually be simulated on the Earth in all it's completeness. The low gravity can be accomplished by flying in a jet at just the right arc to produce 20 percent gravity. The atmosphere is mostly methane with no oxygen so you would have to hold your breath and the temperature is about -150 F which is barely tolerable so you might want to bring a coat. All we would have to do is build a 360 degree projection screen in the jet of the landscape and we are there … for 35 seconds of the jet arc.

But hey maybe we can just go and terraform Mars to live on. One of the things that experienced scientists never seem to consider when making statements that invite humanity to build cities and live on places like Mars, is the economics. Mars is a disaster of a planet, an abused child so to speak. It has done nothing but take it in the ass from Jupiter since the beginning of time so to speak. It exists in that special part of the solar system that doesn't allow for a whole lot of consistency in terms of building a strong healthy planet. It is the fourth planet out with Jupiter as the fifth planet. The tattered asteroid belt that lies in between the two offers a clue. Normally the asteroid belt might have formed a normal planet, but hanging out next to the massive Jupiter doesn't make that possible. This alone is a recipe for eternal disaster because every time Jupiter passes by in a close approach it uses it's massive gravity to rip the hell out of Mars. As a result Mars has formed with a stunted growth resulting in only 38 percent Earth gravity, no magnetic field and a bullshit C02 atmosphere about 1 percent that of Earth pressure. This basically means that anyone who decides to live on Mars will suffer serious muscle degradation unless ,of course, they are willing to spend a good chunk of their lives exercising to maintain normal earth evolved muscle tone. Yes, we will go there in small numbers and probably mine the shit out of it. I mean, after all, why should Jupiter have all the fun.

Prologue

2001: A life Odyssey

As a old child I was deeply affected by the Ape/monolith scene in the film 2001: A Space Odyssey. It opened up areas of the brain that would have normally never been accessed. The experience eventually grew and became so powerful in my mind that I actually consider it on par with an actual extraterrestrial experience. I was only five at the time, yet I understood that a strange and bizarre logic was occurring. It began with the 'dawn of man' sequence 4 1/2 million years ago when a primitive tribe of Australopithecus, whom at that point in their existence had only experienced the natural formations of nature, now suddenly were experiencing the strange shape of a perfectly rectangular black slab amidst the natural jumble of nature. It was a moment so haunting and so freaky I often wonder how human beings thought of it. It was so powerful because within our sub conscious minds we detected some sort of strange evolutionary occurrence may be taking place. What? We do not know; only that we feel we are looking at something that may have occurred in nature beyond our natural experience. That and a little acid in 1987 made me the man I am today!
What the monolith's occurrence ultimately inspired must be viewed in the film for a more proper cosmic experience *(I recommend only a half of hit of acid. Consult your physician)*. And it is one of the few films which allows a lifetime of growth of imagination beyond the film itself. This is because it

is presented in such perfect mystery, never showing us too much and most importantly of all, never really explaining it's own extraterrestrial encounter. This is what a true alien experience would be like. A complete lack of understanding of what we are seeing; the monolith defines that. Where did it come from? What put it there? Could the monolith have been a small piece of a consciousness spanning across the universe? Don't answer that! I don't want to know. Could it be a central consciousness seated deep within the depths of space, compiling knowledge that is incoming from all directions at the speed of light from millions of light years around? Again: I don't want to know and I do want to know. But I'm in love with the mystery. It has created that wonderful sense of awe that now allows me to look at a blue sky with a fusion sun burning brightly overhead and become engulfed in cosmic awe, which is now on par with sex and laughter as pleasurable brain euphoria.

Chapter Index

WELCOME TO THE MIDDLE OF INFINITY

MASS COMMUNICATION BIRTH

EVOLUTION THEORY GONE OVERBOARD

Warning: If you are heavy into religion, viewer discretion is advised!

Fun with Mass Extinction

Intelligent design for the purposes of evolutionary occurrence

An exactitude of the laws of physics should imply an exactitude of the laws of events (see pool table for proof below).

Time for more acid

Everything is a perfect storm

THE POSSIBILITY OF ACTUALLY PARTYING WITH ALIENS

Universe

Dark energy must die because it doesn't exist

Time travel without pissing off any Gods

What if the speed of light and 186,282 miles per second are, in fact, not the same thing. An alternative theory of traveling at the speed of light.

Other ways to 86 this dimension

You can't get there from here

The effects of Atomic Substance Theory on the human brain

Google Milky Way

How slow the speed of light is

WHAT GOES UP NEVER COMES DOWN, SPINNING WHEELS...

In space no one can hear you scream for customer service that is;

Field mice and the center of the galaxy

Digital record erosion

Life in another dimension

THE NON-TECHNOLOGICAL GODS

Driving Miss Dolphin

Dolphin brain reality

The mysteries that dolphins must contemplate

Dolphin addiction

Life is a ballet

The dolphin erotic

The Dolphin P.H.D.

The digital dolphin of tomorrow

The dolphin/Europa connection

Dolphins and the Center of the Galaxy

Why Europa could be a land planet

Binary star bizarre

Imagining aliens by imagining their environments

The Third Gravity

THE GODS ARE CHASING THEIR ANIMATIONS INTO INFINITY

Flintstones; meet the Flintstones

"If it ain't broke don't fix it!

TOMORROW NEVER DIES

Welcome to my new super number:

Why do objects attract?

Pool exactitude

Infinity thinks its cool just because its infinite

Are we happening elsewhere

Finding the Millennium Falcon

Will the collision of the Milky Way Galaxy and M31 in Andromeda yield a new kind of galactic bio dynamic

Does any lifeform species survive into infinity

Prologue: 2001: A life Odyssey

Sources:

'Cosmos' Dr. Carl Sagan, time dilation calculations for starship travel as well as inspiration for just about everything in this book.
Sir Arthur Eddingtion
'Moonraker' Director, Lewis Gilbert
'Octopussy' Director, John Glen
'Never Say Never Again' Director, Irvin Kershner
'Thunderball' Director, Terence Young
Clint Eastwood
Animator, Chuck Jones
Warner Brothers, Bugs Bunny
William Shatner
Star Trek, Creator, Gene Roddenberry
Gilligan's Island, Creator, Sherwood Schwartz
Communion: Author, Whitley Strieber
Google Earth
Animator, Tex Avery
'2001: A Space Odyssey', Metro Golden Mayer
Director: Stanley Kubrick, Writer: Arthur C. Clark
Star Wars: A New Hope', Twentieth Century Fox,
Writer/Director: George Lucas

About the Author

Paul Vancleve is a Blues guitarist, cosmic theoretician and a few other things that can't be mentioned here.

Learn more about Paul Vancleve on either his *Andromeda kid'* channel or *'Paul Vancleve'* channel on Youtube through the videos:

Learn more about Paul Vancleve on either his *Andromeda kid'* channel or *'Paul Vancleve'* channel on Youtube through the videos:

Paul Vancleve is Andromeda Kid

Song/video '*Ode to thy 45th Potus*' on Youtube

Song/video '*Operation Naughty*' on Youtube

Buttdials from the Center of the Galaxy (21 videos)

'The Dirty Harry Scriptures'

Trump: Double Agent working for the Democratic Party'

'Trump: PDB Follies'

'Trump Hypnotized'

Other Books by Paul Vancleve:

'Butt dials from the Center of the Galaxy'

And

'Kinetic Evolution as Interpreted by a Pop Cultural Consciousness at Mass Communication Birth Super

hell Second'

Music by Paul Vancleve

Songs:

'Operation Naughty'

'Rock that Naughty'

'The Search for the Ancient Cosmic Background Porn'

'Butterfly Pyramids'

'Crushdepth'

'Ode to Thy 45th Potus'

'Hyme to Thy Neightborhood Pyro'

Later Darth Vader'

'Pound Me'

Blues with Orgasmic Intent

'Blues for the Criminally Insane'

'Insanity Blues for Serial Killer Cannibals'

'Blues for Alien with Overbooked Flights'

'Psychotic Blues for Disturbed Aliens'

'Tornado Blues for People in Tornados'

'Orgy Blues for Proxima Centauri B'

'Fractal Blues Upon a Hopeful Mind'

'714's Guide For a New Generation'

Ode to Witch Hazel

All Available on all major Download Services

Contact Info:

Twitter: *andromeda_kid*
Instagram: kidandromeda
Facebook: *Paul Vancleve*

Email: paulvancleve@yahoo.com
This is where to reach me.

This book is also available on Amazon/Kindle with color imagery